Fascinating Life Sciences

This interdisciplinary series brings together the most essential and captivating topics in the life sciences. They range from the plant sciences to zoology, from the microbiome to macrobiome, and from basic biology to biotechnology. The series not only highlights fascinating research; it also discusses major challenges associated with the life sciences and related disciplines and outlines future research directions. Individual volumes provide in-depth information, are richly illustrated with photographs, illustrations, and maps, and feature suggestions for further reading or glossaries where appropriate.

Interested researchers in all areas of the life sciences, as well as biology enthusiasts, will find the series' interdisciplinary focus and highly readable volumes especially appealing.

George Poinar

Flowers in Amber

 Springer

George Poinar
Department of Integrative Biology
Oregon State University
Corvallis, OR, USA

ISSN 2509-6745   ISSN 2509-6753   (electronic)
Fascinating Life Sciences
ISBN 978-3-031-09046-2   ISBN 978-3-031-09044-8   (eBook)
https://doi.org/10.1007/978-3-031-09044-8

Cover image: In this beautiful mid-Cretaceous flower of Valviloculus pleristaminis, some 50 stamens positioned on top of a receptacle are shedding their pollen.

This Springer imprint is published by the registered company Springer Nature Switzerland AG
The registered company address is: Gewerbestrasse 11, 6330 Cham, Switzerland

Kenton L. Chambers (April, 2010)

This book is dedicated to my friend and colleague, Kenton L. Chambers, of the Department of Botany and Plant Pathology, Oregon State University in Corvallis. Ken's botanical training and broad experience has provided him with a wide knowledge of plant morphology and systematics. This knowledge has been indispensable in studying amber flowers from around the world.

Introduction

While much attention has been given to animal life in amber, the remains of a variety of vegetation, including angiosperm flowers, also exist in fossilized resin. The present work is a pictorial synopsis of 94 flowers and their pollinators and herbivores that occur in 4 major amber deposits around the world [Burmese (Myanmar), Baltic, Dominican Republic, and Mexican]. It is written for an informed popular audience with emphasis on the beauty, diversity, and curious features of ancient flowers in amber. This work substantially presents new information, including interesting facts about the likely habits and natural history of the plants that bore these amber flowers, as well as a synopsis of their major structural features. Tables with flowers from the four amber deposits with their families and reference to technical descriptions. New included topics cover pollinators and herbivores associated with these flowers. The inclusion of keys to the identification of these fossil flowers will be appreciated by those interested in making comparisons with new material.

The diverse flowers in mid-Cretaceous Burmese amber existed at a crucial period of angiosperm evolution and most are unlike any flowers present today. With flowers in Baltic amber, roughly half can be assigned to modern families, while flowers in Dominican and Mexican amber can usually be placed in extant families and often in current genera. Comments on foliar and growth characteristics of the plants that bore these flowers are based on features of their closest present-day descendants.

As the diversity of angiosperms increased, their associated insects proliferated. The flowers responded by producing additional scents, colorful petals, and more nectar to attract a wide range of pollinators, especially beetles, flies, wasps, and bees. These changes are evident as we pass from mid-Cretaceous to mid-Tertiary flowers.

Our ability to view flowers from these ancient landscapes is due to a few tree species that produced resin that persisted for millions of years. In the Burmese and Baltic amber forests, it was mainly kauri trees (Araucariaceae) that served this purpose, while in the mid-Tertiary Dominican Republic and Mexican amber forests, *Hymenaea* legume trees supplied the resin.

The Miracle of Amber

Flowers, as well as leaves, falling into tree resin show an amazing degree of preservation, obtained in part by chemicals in the resin that removed water from the plant's tissues and acted as natural embalming agents.

Since most plant resins quickly deteriorate, it is only those from a few trees that persist and harden into amber. When the resin begins to polymerize, volatile terpenes escape, and nonvolatile terpenes bind together, forming compounds resistant to natural decay processes. The enclosed flowers can then be studied, ancient landscapes reconstructed, and we are presented with a time capsule from millions of years ago. Fortunately, amber is soft enough to be further processed. In order to observe fine details of the flowers, most amber has to be cut, reshaped, and then polished. Afterwards, the flowers are orientated on slides for viewing at different light intensities under both binocular and compound microscopes. New methods, such as micro-CT scans that provide a non-destructive view of internal tissues, are being used more frequently in the study of fossil flowers.

Floral Characteristics

Angiosperm flowers typically consist of a central pistil, a ring of stamens, an inner circle of petals, an outer ring of sepals, and a basal stalk (pedicel). Sometimes if there is only one circle of structures that are not clearly sepals or petals, the term "tepal" is used to describe them. Flowers bearing both stamens and pistils are perfect or bisexual species. Those with only stamens or only pistils are termed imperfect, unisexual, or just male and female flowers. The pistil, or female part of a flower, includes an ovary with a terminal style capped by a stigma. A number of female flowers contain sterile stamens known as staminodes. These sterile male structures are normally reduced and consist of filaments without functional anthers. Stamens, or the male parts of a flower, consist of a filament supporting an anther that produces pollen. Some male flowers possess undeveloped female structures called pistillodes. These sterile female structures are often reduced in size and are not functional. Flowers in full bloom are said to be in anthesis.

Pollination occurs when pollen grains are transported from an anther to a stigma. Fertilization occurs when pollen grains on the stigma form pollen tubes that enter and grow down the style to fertilize eggs in the ovary. It is interesting that stages of the pollination process were captured on some of the Burmese amber flowers.

Acknowledgments

I gratefully acknowledge the love, assistance, support, and understanding of my wife and coworker, Roberta, who was by my side all through the preparation of this work. Without her encouragement and discussions, this work would never have been published.

Flowers in amber were collected over many years at various gem trade shows, especially in Tuscon, Arizona, as well as from miners in the Dominican Republic and Chiapas, Mexico. I would like to thank the University of California Research Expeditions Program (UREP) for support, allowing my wife and I to lead an expedition to study amber mines and obtain amber in the Dominican Republic in 1987. I would also like to thank the late Lauren Faunt and Renato Zárate for inviting me to their laboratory in Chiapas in 1982 and taking me to the Mexican amber mines in the vicinity of Simohovel. Others who have supplied me with flowers in amber over the years are Ron Buckley, Jim Work, Alex E. Brown, Fernando E. Vega, Andrew Cholewinski, Ramón Martinez, Sieghard Ellenberger, and the late Dan McAuley, Aldo Costa, and Jake Brodzinsky.

All photos were taken by the present author with the exception of Fig. 1.1 in Chap. 1 that was supplied by Sieghard Ellenberger and Fig. 4.1 in Chap. 4 that was taken by the late Wyatt Durham. I thank Barney Lipscome for permission to use photos previously published in the *Journal of the Botanical Research Institute of Texas* in the present work. I am also grateful to the late Mary White for discussions on primitive angiosperms now surviving in refugia in Australia and the possible origin of angiosperms in Gondwana.

Contents

Chapter 1
Burmese Amber Flowers

Abstract The 31 species of flowers that have been recovered from Burmese amber shows the amazing variety of angiosperms that were present in this mid-Cretaceous forest that flourished during the reign of the dinosaurs. Based on their floral diversity, it is obvious that this was a period of experimentation with only a few angiosperm lineages, such as the grasses and laurels, carried through to the present. In amber from this deposit, you can find one flower with multiple sized sepals and other flower with secondary anthers developing from the backs of primary anthers. Angiosperms were also "experimenting" with different ways to attract pollinators and deter herbivores as part of their adaptation to various habitats. Some of these fossil flowers were instrumental in establishing the theory that the flora and fauna of Burmese amber was formed in Gondwana and later rafted to its present location in northern Myanmar.

The mid-Cretaceous was the dawn of angiosperms. Amidst the dominant archaic conifers, ginkgoes, cycads and horsetails, angiosperms were beginning to diversify some 100 million years ago (Cruickshank and Ko 2003; Shi et al. 2012). Most of these primitive flowers were quite small and many lacked petals. Small flowers are also characteristic of many flowers growing in today's heavily shaded tropical forests. Most blooms in the Burmese amber forest probably came from shrubs and small trees, although some could have been from vines and others could have been attached to the cylindrical buttresses or furrowed stems of gigantic trees. A few may have grown on stumps, trunks and branches, together with ferns, moss and lichens. Other microhabitats could have been under the protection of tree ferns that reached up to 15 feet in height or as epiphytes hanging from the branches of Kauri trees. Instead of using brightly colored petals to attract insects, these primitive flowers offered rewards of pollen, nectar and odors to attract potential pollinators. Like many flowers in today's tropical rainforests, the inconspicuous Burmese amber flowers were probably quickly deciduous, opening in the morning and falling in the evening or the following day. Thus the chances of catching mature flowers in resin would be slight. However, based on insect damage on many of the flowers, it appears that some disturbance, such as visiting herbivores, detached many flowers when

© The Author(s), under exclusive license to Springer Nature Switzerland AG 2022
G. Poinar, *Flowers in Amber*, Fascinating Life Sciences,
https://doi.org/10.1007/978-3-031-09044-8_1

Fig. 1.1 Miners digging and sorting Burmese amber in the Hukawng Valley of Kachin State. (Photo courtesy of Sieghard Ellenberger)

they were still in their prime or just beginning to wane. Sometimes the perpetrators were preserved along with the flowers. While the damage to flowers may alter their appearance, it normally does not prevent their description.

Amber from Burma (Myanmar) is collected from simple mines dug in the ground or into the sides of mountains (Fig. 1.1). Evidence from flowers suggests that the Burmese amber forest was originally part of Gondwana, quite some distance from where the amber is now located in northern Myanmar. At the outset, it was attached to the combined terrains of Australia, Africa, South America, New Zealand, Antarctica and the Indian subcontinent. When the amber forest separated from the rest of Gondwana, it carried a distinct selection of plants and animals that had already been entombed. This explains why many flowering plants in Burmese amber have descendants now in countries that originally were part of Gondwana (Poinar 2018a).

1.1 Mid-Cretaceous Flowers

It was fortunate that Kauri trees (*Agathis* sp.: Araucariaceae) were present in the ancient Burmese rainforests since it was their resin that preserved the flowers (Poinar et al. 2007b). In New Zealand, it is still possible to find majestic Kauri trees in small, isolated forests (Fig. 1.2). Resin exuding from the limbs and especially the main trunk, often reaches the forest floor (Fig. 1.3). These trees, which extend up into the canopy and live for 1000 years, have wide and glossy leaves (Fig. 1.4). Their round seed cones are very hard (Fig. 1.5) but the scales fall apart readily at maturity.

Large areas of the trunks and limbs of the Kauri trees in the Burmese amber forest would have been covered with epiphytes, which may explain tips of fern leaves appearing in Burmese amber (Fig. 1.6). However, some of these "epiphytes" may actually be camouflaged beetles that bear dorsal protrusions disguising them as moss (Fig. 1.7).

Valviloculus pleristaminis, description based on Poinar et al. 2020

Like so many ancient flowers in amber, this small 2.7 mm wide, male *Valviloculus pleristaminis* with its large floral cup and numerous (30+) upward pointing stamens, is beautifully preserved. The 6 enclosing sepals curve upwards and appear to be protecting the anthers (Fig. 1.8).

Fig. 1.2   Extant Kauri tree in New Zealand

Fig. 1.3 Resin exuded
from trunk of extant
Kauri tree

Fig. 1.4 Leaves of extant Kauri tree

Fig. 1.5 Extant Kauri seed cones

By itself, a single *Valviloculus* flower would never be noticed, but it was probably one of many, all clustered together in a large flower head, perhaps with accompanying female flowers. The completely exposed numerous anthers of this fossil have released golden colored pollen.

Although *Valviloculus* contains only productive stamens (Fig. 1.9), in the center are a number of undeveloped female styles (pistillodes). These sterile female styles indicate that originally, *Valviloculus* had both functional stamens and pistils (bisexual). Having male and female sexes on different plants would appear to be counterproductive, since the pollen would have to travel between two flowers for seeds to be formed. But this condition is fairly common today and perhaps favors cross pollination, thus providing additional genetic diversity to the species. Beneath the stamens on the flower is a large floral cup that is completely hollow (Fig. 1.8) but would have contained developing seeds in female flowers.

Pollen from *Valviloculus* would have attracted beetles, flies and other small insects. Appealing odors were probably emitted from glands at the tips of the anthers and if the fruits of *Valviloculus* were fleshy, like some cherries or plums, they would have attracted a number of breeding insects. Some small vertebrates also may have been able to consume the entire inflorescence.

Searching for present-day flowers with a similar structure takes us to New Zealand where the Pigeonwood tree thrives (*Hedycayra arborea*: Monimiaceae). This attractive tree grows to a height of 45 feet and has glossy, dark green leaves. As

Fig. 1.6 Tip of fern, *Cladarastega burmanica* Poinar (2021a), in Burmese amber

with *Valviloculus*, male and female Pigeonwood flowers are borne on separate trees, the male flowers are surrounded by sepals and the stamens are all crowded together at the top of the flower. The fruits of *Hedycayra arborea* are fleshy drupes similar to cherries.

Dispariflora robertae, description based on Poinar and Chambers 2019c

Dispariflora robertae is one of the most unique flowers in Burmese amber. Not only does *Dispariflora* have different sized sepals, but new flowers develop from lateral branches of old flowers. Here are three interconnected *Dispariflora robertae* flowers in a single piece of amber (Fig. 1.10). In life, the interconnected flowers could have formed quite a large flower head, possible extending over most of the plant. The flowers have a single bristly ovary with a very short style capped by a swollen stigma. As many as 15 stamens, each with basal secretory glands to attract pollinators, are present.

Fig. 1.7 Camouflaged beetle, *Stegastochlidus saraemchena* Poinar and Vega (2020), in Burmese amber

All flowers are bisexual and while they lack petals, each flower possessed a circle of sepals that vary greatly in size. In fact, the largest sepal, which is 9.0 mm in length, could easily pass for a leaf, being almost 3½ times longer then the next largest sepal.

How sepals evolved in angiosperms is unknown. Since most sepals contain chloroplasts and have leaf-like features, it is assumed that they developed directly from leaves. *Dispariflora robertae* may have been in the process of forming sepals through the reduction of its leaves (Fig. 1.11). How long would it take all of the surrounding sepals to reach the size of the smallest one: weeks, months, generations? This is one of the many Burmese amber flowers that could not be assigned to a modern family and there are no living flowers that have such a range of sepal size.

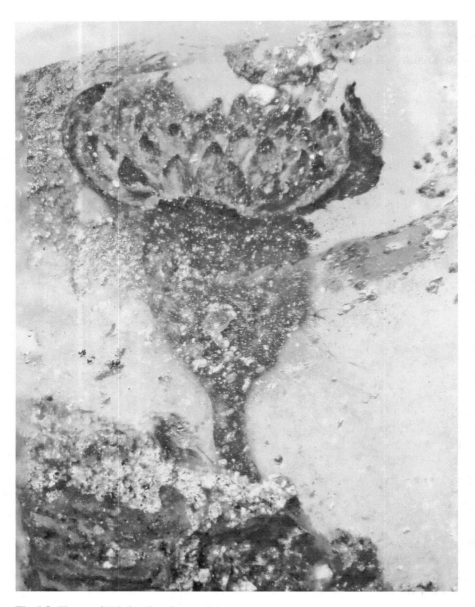

Fig. 1.8 Flower of *Valviloculus pleristaminis*

Small bees and various herbivorous insects could have served as pollinators while they devoured or collected pollen. The remains of two arthropods are entombed with *Dispariflora robertae*. These include a solitary insect leg and 8 detached spider legs. Obviously a predator had attacked these floral visitors and after consuming their bodies, discarded the legs. This disturbance quite likely

Fig. 1.9 Stamens of *Valviloculus pleristaminis*

Fig. 1.10 Three interconnected *Dispariflora robertae* flowers

Fig. 1.11 Single *Dispariflora robertae* flower

caused the flower to fall. Aside from the variable sizes of the sepals, features of *Dispariflora* show affinities with present day members of the laurels.

Tropidogyne pikea, description based on Chambers et al. 2010; ***Tropidogyne pentaptera***, description based on Poinar and Chambers 2017; ***Tropidogyne lobodisca***, description based on Poinar and Chambers 2019a; ***Tropidogyne euthystyla***, description from Poinar et al. 2021

All four species of *Tropidogyne* that flourished in the Burmese amber forest are similar in lacking petals, but having 5 sepals of approximately equal length with 1–5 major veins and anastomosing veinlets (Fig. 1.12). One flower, considered a mutant, had 6 sepals (Fig. 1.13). What distinguished the separate species is their size (from 2.0 to 5.1 mm in width), sexual status (unisexual or bisexual), number of stamens (0, 5 or 10), form of their ovaries and shape of their styles (2, 3, long, short, erect, arching).

Since brightly colored petals, which are used today to attract pollinators, were rare on mid-Cretaceous flowers, scents, nectar and other glandular secretions were used to attract pollinators. All *Tropidogyne* species possess nectaries and glandular

Fig. 1.12   *Tropidogyne pentaptera* flower with 5 sepals

Fig. 1.13   *Tropidogyne pentaptera* flower with 6 sepals

hairs on the margins of their sepals. These nectaries produced fragrances and other sweet deposits while the glandular hairs secreted oil droplets (Fig. 1.14) (Poinar and Poinar 2018). These deposits would have been quite attractive to a number of insects, including primitive bees like *Discoscapa apicula* (Poinar 2020) (Fig. 1.15) and *Melittosphex burmensis* (Poinar and Danforth 2006; Danforth and Poinar 2011) (Fig. 1.16) as well as Burmese amber flower beetles (*Furcalabratum burmanicum* Poinar and Brown 2018) (Fig. 1.17). While searching for nectar and collecting pollen, these insects would have inadvertently pollinated the flowers.

Tropidogyne is one of the most common flower genera in Burmese amber. Perhaps their favorite habitat was in the shade of kauri trees, thus increasing their changes of ending up in amber. *Tropidogyne* species are related to coachwood trees of the present day genus *Ceratopetalum* in the family Cunoniaceae.

Fig. 1.14 Oil droplets released from *Tropidogyne*

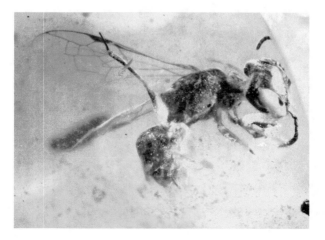

Fig. 1.15 Female bee (*Discoscapa apicula*)

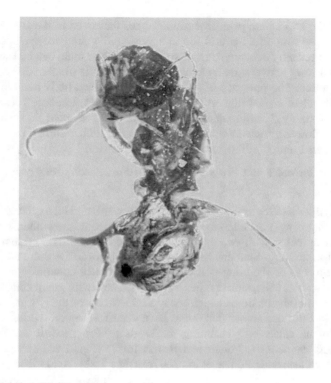

Fig. 1.16 Male bee (*Melittosphex burmensis*)

Fig. 1.17 Flower beetle (*Furcalabratum burmanicum*)

Both of the two small, primitive solitary bees that lived in the Burmese amber forest had some structural features found in present day predatory wasps, revealing that they had recently evolved from wasp ancestors. The male bee had some pollen grains still resting between its branched body hairs while the female bee was covered with pollen. The presence of pollen among branched body hairs in both bees show that they had visited flowers just before they were entombed. These "minute" bees (both roughly 3 millimeters in length) are well adapted for visiting small angiosperm flowers, such as those in the Burmese amber forest, most of which range between 1 and 5 millimeters in greatest dimension.

Programinis burmitis and *Programinis laminatus*, descriptions based on Poinar 2004, 2011

The description of the grass genus *Programinis* was based on a spikelet (Fig. 1.18) and a leaf fragment (Fig. 1.19) in separate pieces of Burmese amber. The laterally compressed spikelet, with two sterile glumes and a series of 5 or more bisexual florets, is 14 mm long. Stamens have emerged from several florets and the anthers are in the process of shedding pollen. The leaf fragment contains epidermal cells with long and short cells, silica bodies, rows of stomata with guard cells (Fig. 1.20), papillae and microhairs, all of which are typical features of today's grasses.

Insects in Burmese amber that could have been feeding on the leaves, sucking juices from the canes or consuming the seeds of *Programinis* are leafhoppers (*Priscacutius denticularis* Poinar and Brown 2017) (Fig. 1.21) and grasshoppers (*Longioculus burmensis* Poinar et al. 2007c) (Fig. 1.22). Aside from insects, the developing seeds of *Programinis* spp. were attacked by an ergot fungus (*Palaeoclaviceps parasiticus*: Clavicipitaceae) (Poinar et al. 2015). The erect, black sclerotium of this ergot, which consisted of fungal hyphae and spores, was protruding from the tips of the florets (Fig. 1.23). Extant ergot-infected grass florets are poisonous to a range of mammals and birds and it is unknown if dinosaurs that fed on infected *Programinis* spp. would have been affected.

Grasses are one of the most successful families of flowering plants, with over 10,000 species living in every habitat imaginable. *Programinis* is considered to be a type of bamboo, a lineage of primitive grasses with their origin occurring in tropical forests, similar to the habitat proposed for the Burmese amber forest. The grass family (Poaceae) is one of the few Burmese amber angiosperm lineages that still survives today.

Dasykothon leptomiscus, description based on Poinar and Chambers 2020c

The flower of *Dasykothon leptomiscus*, with its 6.7 mm long pedicel, is unique in several ways. Aside from its extremely long pedice (Fig. 1.24) and 2 elongate curved styles that protrude from opposite sides of the flower (Fig. 1.25), it has two types of sepals; 3 of which are oval and cup shaped, while the other two are narrow. The anthers are missing on most of the 12 stamens but there is an abundant supply of pollen on one of the five remaining anthers. The pollen grains have a curious shell-like shape and attach to the stigma by their tips and edges. Some pollen grains had recently germinated and it was possible to see the pollen tubes entering the

Fig. 1.18 *Programinis burmitis* spikelet

stigma and growing down the style towards the ovary (Fig. 1.26). While differing from any modern plant family, the pollen of *Dasykothon leptomiscus* has some features found in representatives of the Southern Hemisphere family Atherospermataceae.

Zygadelphus aetheus, description based on Poinar and Chambers 2019d

Zygadelphus aetheus is another bisexual flower lacking petals but with 10 spirally arranged sepals ranging from 0.75 to 2.0 mm in length that partially enclose the stamens and two styles (Fig. 1.27). The most incredulous feature of *Zygadelphus aetheus* is its ability to produce secondary anthers from the tips of primary anthers

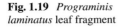

Fig. 1.19 *Programinis*
laminatus leaf fragment

(Fig. 1.28). This explains why the flower contains only 4 stamens, but 8 anthers. These piggyback secondary anthers are functional and produce abundant spherical pollen grains similar in form to those of the primary anthers. The stigmas are positioned in such a way to be dusted by pollen from the secondary anthers. This distinctive feature of anther duplication could not be found on any other living or extinct angiosperms. While placement of *Zygadelphus* within a modern family is not possible, especially regarding the type of secondary anther development, features of the 4 primary stamens are shared with members of the pan-tropical family Hernandiaceae. Growth patterns in this family include trees, shrubs and vines, with pollination mainly by insects.

Lachnociona terriae, description based on Poinar et al. 2008

Lachnociona terriae represents a hirsute, functionally pistillate flower with 5 strongly curved hirsute sepals, no petals, 10 staminodial filaments and a 5-carpellate hirsute columnar pistil with 5 stout connate styles with recurved tips (Fig. 1.29).

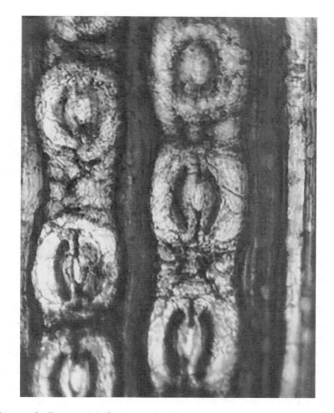

Fig. 1.20 Stomata in *Programinis laminatus* leaf fragment

Fig. 1.21 Leafhopper in Burmese amber

Fig. 1.22 Grasshopper in Burmese amber

Some of the slender filaments of the staminodes lack functional anthers and in their place bear simple vestigial anthers that are little more than undifferentiated tissue lacking cellular structure (Fig. 1.30).

While the flower is one of the larger species recovered from Burmese amber (5.6 mm in length), the dominant feature of *Lachnociona terriae* is its ovary bearing 5 chimney-like, hirsute styles. *Lachnociona terriae* is functionally female even though it has 10 abortive male stamens (staminodes). These staminodes suggests that the precursor of *Lachnociona terriae* was bisexual and at this time, the flowers were becoming unisexual. Having an arrangement of unisexual rather than bisexual flowers would increase the chances of cross pollination and producing seeds with "hybrid vigor".

The other unusual feature of *Lachnociona terriae* is the long twisted hairs (trichomes) on the surface of the sepals. Perhaps they served to protect the sepals against herbivorous insects since while insect damage was evident on the staminodes, none was noted on the sepals. Types of pollinating insects for this flower are unknown and it was not possible to determine if the flower had a nectar disc. However, it is possible that the flower produced other odoriferous scents to attract specific pollinators.

Many characters of *Lachnociona terriae* resemble those found in members of the present day Southern Hemisphere families Brunelliaceae and Cunoniaceae. Plants in these families are mostly shrubs and trees with leathery leaves and small, white flowers that produce nectar. Pollination is by insects, birds, lizards and bats.

Lachnociona camptostylus, description based on Poinar and Chambers 2018c

Two flowers entombed together in a single piece of Burmese amber with features of *Lachnociona terriae* were placed in the same genus but described as *Lachnociona*

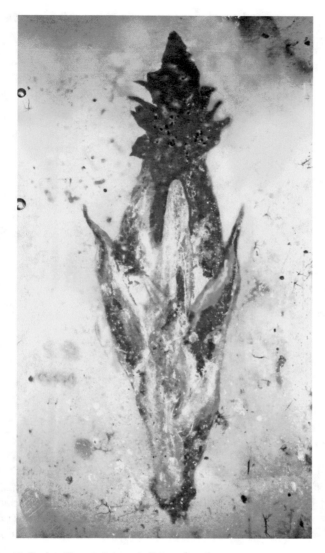

Fig. 1.23 Ergot infecting *Programinis* sp. in Burmese amber

camptostylus (Fig. 1.31). Differences between the two flowers of *Lachnociona camptostylus* are considered to represent variation within a single species. Both flowers are only 2.4 millimeters long, with 5 large sepals, no petals and two back-ward diverging separate styles. One of the flowers is functionally pistillate with 5 styles, no nectar glands and three staminodes (Fig. 1.32). Odors or volatile oils were probably released from the pistillate flower to attract pollinators. The other flower is hermaphroditic (bisexual) and possesses well developed functional stamens and 4

Fig. 1.24 *Dasykothon leptomiscus* with long pedicel

styles (Fig. 1.33). Ten large, pebbly nectar glands form a ring around the base of the styles. Pollen grains of *Lachnociona camptostylus* are tricolpate (Fig. 1.34).

It is not possible to assign *Lachnociona camptostylus* to an extant family, although as with *Lachnociona terriae*, the flowers have some features of the primitive Australian Cunoniaceae and Central and South American Brunelliaceae.

Setitheca lativalva, description based on Poinar and Chambers 2018b

Setitheca lativalva is a staminate flower measuring 4.9 mm across the outstretched tepals (Fig. 1.35). Petals are lacking and the edge of the flower displays a series of 12 spirally arranged flat tepals. The tepals are of varying sizes and shapes with irregular margins lined with fine, short hairs, some of which appear to be emitting microscopic droplets, either to repel herbivores or attract pollinators. The 10 stamens, which are attached to the margin of a flat central disc, have striking dark colored anthers.

The stamens look like miniature mushrooms (Fig. 1.36) and have some interesting features. At the base of the filaments are 2 slender glands that probably emit nectar to attract pollinators. The anthers are bilocular with each lobe containing a

Fig. 1.25   *Dasykothon leptomiscus* flower

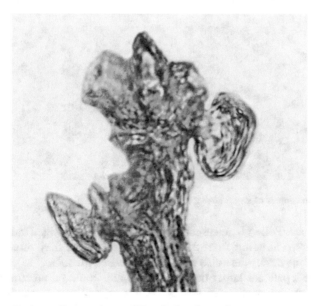

Fig. 1.26   Germinating pollen on stigma of *Dasykothon leptomiscus*

Fig. 1.27 Flower of *Zygadelphus aetheus*

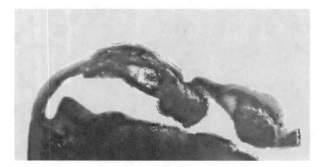

Fig. 1.28 Second anther of *Zygadelphus aetheus* growing out of first anther

single pollen sac. Pollen is released via 2 lateral, dorsally-hinged valves. The stiff, spine-like hairs protruding from the anther valves, which may protect the pollen from herbivorous insects, are also unique to *Setitheca lativalva*.

Five of the tepals are larger than the others and could be precursors of petals, which brings up the question of whether petals or sepals appeared first on angiosperm flowers. Since sepals enclose and protect the flower bud before opening and

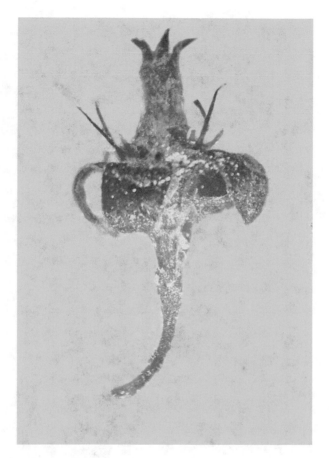

Fig. 1.29 *Lachnociona terriae* flower

are the first part of the flower to develop, they probably evolved before petals. Many of the Burmese amber flowers posses sepals, but lack petals, whose main function is to attract specific pollinators with colors and aromatic compounds.

Palaeoanthella huangii, description based on Poinar and Chambers 2005

Palaeoanthella huangii is one of the smallest flowers studied in Burmese amber, being only 1 millimeter in diameter (Fig. 1.37). The flat, round staminate flower has a cup shaped perianth composed of 8 fused tepals that are separate above but connate below. There are no petals or nectaries. The tepals alternate with a ring of 8 sub-sessile 2-lobed stamens. The pollen is circular, inaperturate, pitted with finely ridged and grooved exines (Fig. 1.38). Features of *Palaeoanthella huangii*, especially its short stamen filaments, resemble members of the extant family Monimiaceae. Gall midges are known to deposit eggs in male flowers of Monimiaceae and a gall midge whose larvae could have caused the damage to the

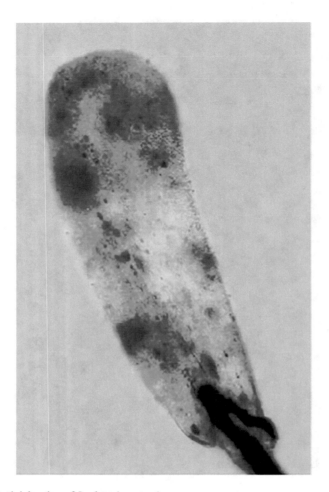

Fig. 1.30 Vestigial anther of *Lachnociona terriae*

tepals of *Palaeoanthella huangi* is entombed adjacent to the flower (Fig. 1.39). Thrips are known to pollinate and raise their brood in flowers of Monimiaceae and these small insects occur in Burmese amber (Fig. 1.40). They would have made excellent pollinators.

Thymolepis toxandra, description based on Chambers and Poinar 2020

Thymolepis toxandra provides an opportunity to examine a past association between insects and flowers. The scene depicts a parasitic wasp (*Paleosyncrasis hongi* Poinar 2019a) that was visiting a flower of *Thymolepis toxandra* being consumed by insect herbivores (Fig. 1.41). The wasp, with her long ovipositor, may have visited the flower to deposit eggs in the develping larvae or have been

Fig. 1.31 Two flowers of
Lachnociona camptostylus

searching for nectar and pollen. Especially curious were several small insect larvae
attached to the hind leg of the fleeing wasp.

Thymolepis toxandra is unique in being a fairly large bisexual flower slightly
over 6 millimeters wide with a series of 12 variously sized tepals arranged in opposite pairs at several levels (Fig. 1.42). The surfaces of the larger sepals are covered
with papillae producing glandular secretions. These secretions may have been to
deter insect herbivores since no damage was evident on these tepals while insect
larvae were feeding on the smaller glandless tepals.

The other unique feature of *Thymolepis toxandra* is the presence of only two
stamens. Both stamens are quite compact with a distinctly curved short filament.
The horseshoe-shaped anthers are covered with a thick layer of hairs (Fig. 1.43). A
bilobed stigma is present at the tip of the fused styles. Features of the flower suggest

Fig. 1.32 Pistillate flower of *Lachnociona camptostylus*

that it could be an early representative of the primitive Southern hemisphere family Monimiaceae.

Micropetasos burmensis, description based on Poinar et al. 2013

The inflorescence of *Micropetasos burmensis* preserved in Burmese amber has 18 min bisexual flowers, all under 1 millimeter in width (Fig. 1.44). These flowers possess prominent, curved, hook-like styles, reminiscent of hay bale hooks (Fig. 1.45). Like the majority of other flowers in the Burmese amber forest, *Micropetasos burmensis* lacks any evidence of petals but has 5 irregular sepals. These sepals are brown, similar to the color of the flower stems, but in life they were probably much lighter, possibly even yellow or white and if so, would have functioned as petals to attract pollinators.

All of the flowers contain multiple, minute stamens (up to 60) of varying lengths and tightly clustered around the style, making it difficult to visualize other features of the pistil. The advantage of the curved style, if any, is unknown. It may offer a

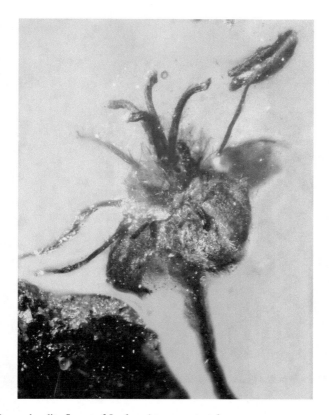

Fig. 1.33 Hermaphrodite flower of *Lachnociona camptostylus*

perch for some specific pollinator, making it easier for pollen to be brushed on the narrow stigma.

Being able to view so many flowers in the same inflorescence provides an opportunity to study floral variation in *Micropetasos burmensis*. For instance, stamen and sepal length differ among the flowers and one flower has a double style. With so many stamens there is an abundance of pollen and in one flower, two oval triaperturate pollen grains had germinated and their pollen tubes had entered the stigma (Fig. 1.46).

In the same amber piece was a scorpion, which raises the question of what possible association could there be between *Micropetasos burmensis* and a predatory arthropod. Since some present day scorpions are known to climb trees, it is possible that this entrapped scorpion was exploring the floral inflorescence when both fell into a pool of resin.

Antiquifloria latifibris, description based on Poinar et al. 2016

Antiquifloria latifibris is a very unique bisexual flower with 16 extended, spirally arranged fleshy sepals that measure 12 mm across the flower (Figs. 1.47 and 1.48).

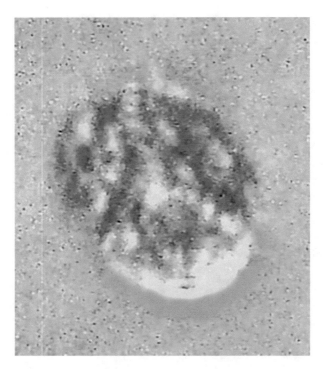

Fig. 1.34 Tricolpate pollen grain of *Lachnociona camptostylus*

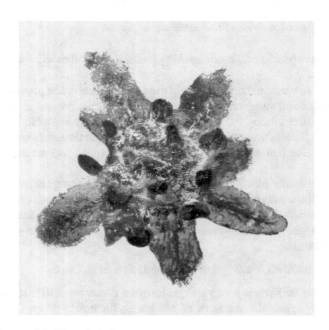

Fig. 1.35 Flower of *Setitheca lativalva*

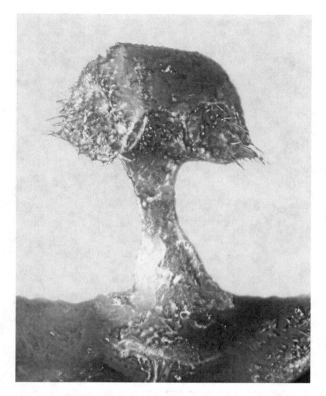

Fig. 1.36   Mushroom-like stamen of *Setitheca lativalva*

The surfaces of these sepals are covered with sharp, pointed hairs and scales that protect the inner core of 12 separate helically arranged stamens and pistil from herbivores. The stamens have their wide, pointed connectives (Fig. 1.49) directed toward the slender stylet with a minute stigma in the center of flower.

At the base of the short filaments is a cluster of parenchyma cells resting on a white pulvinate area. The latter may represent secretions from the parenchyma cells that form a nectiferous ring at the base of the stamens to attract pollinators. A possible pollinator could have been a crane fly that remains adjacent to the flower. Crane flies are known to visit flowers for nectar today and are common in Burmese amber (Fig. 1.50).

The stamens surrounding the style have an interesting type of pollen release. Instead of valves or slits, the pollen is liberated by the deterioration of the walls of the 4 elongate pollen sacs. This releases pollen on floral visitors brushing up against the anthers while feeding on nectar. Based on the present interpretation of its features, *Antiquifloria latifibris* cannot be placed in any modern family.

Fig. 1.37 Flower of *Palaeoanthella huangii*

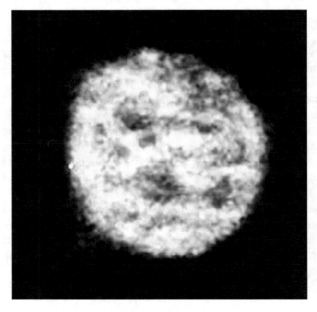

Fig. 1.38 Pollen grain of *Palaeoanthella huangii*

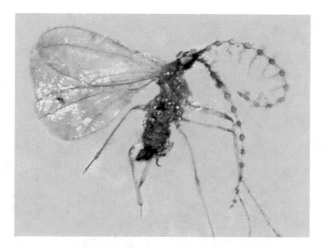

Fig. 1.39 Gall gnat adjacent to *Palaeoanthella huangi*

Fig. 1.40 Thrip in Burmese amber

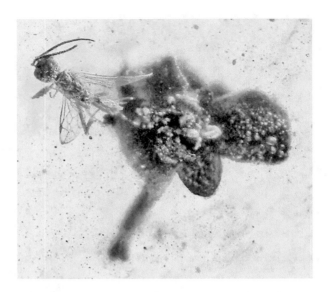

Fig. 1.41 Wasp adjacent to *Thymolepis toxandra*

Fig. 1.42 Flower of *Thymolepis toxandra*

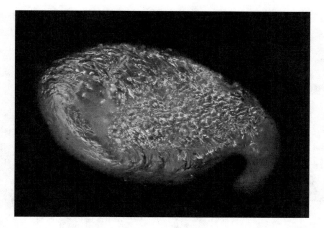

Fig. 1.43 Horseshoe-shaped stamen of *Thymolepis toxandra*

Fig. 1.44 *Micropetasos burmensis* inflorescence

Fig. 1.45   Three flowers of *Micropetasos burmensis*

Cascolaurus burmitis, description based on Poinar 2016

Cascolaurus burmitis is a small unisexual staminate flower just under 6 millimeters in length. Petals are absent however 3 of the 6 sepals are long and curved, resembling reflexed petals (Fig. 1.51). While now brown, when living, the large sepals could have been brighter and played the role of petals by attracting pollinators. Nine erect stamens are positioned in the center of the flower. *Cascolaurus burmitis* represents an extinct member of the laurel family (Lauraceae) and is one of the flowers in the Burmese amber forest that can definitely be assigned to a recent family. Features of *Cascolaurus burmitis* shared today by other members of the Lauraceae are flaps on the anthers that open to release the pollen. In this flower, the anther flaps are fully open and almost all of the pollen has been dispersed (Fig. 1.52).

Cascolaurus burmitis attracts pollinators with six large, nectar glands exposed at the top of the flower. Several of these exposed glands are still covered with secretions that would have attracted a number of insects (Fig. 1.53). Various flies pollinate present day laurel flowers and it is possible that the Burmese amber unicorn fly

Fig. 1.46 Germinating
pollen grains attached to
stigma of *Micropetasos
burmensis*

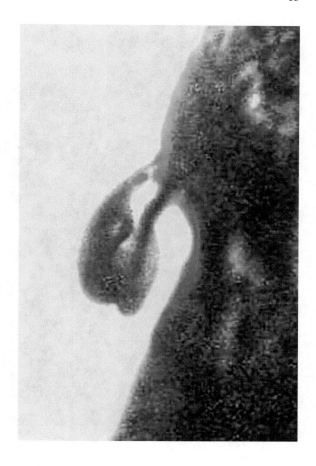

(*Cascoplecia insolitis* Poinar 2010) visited *Cascolaurus burmitis* since with its
reduced mouthparts (Fig. 1.54), the unicorn fly would have had no problem obtaining nectar from the exposed glands. Modern descendants of unicorn flies do visit
flowers.

Flowers of *Cascolaurus burmitis* closely resemble blooms of the present day
genus *Litsea*. Members of this genus occur in tropical and subtropical regions of
India, Africa, Australia, New Zealand and North and Central America. The flowers
of *Litsea* are borne in small clusters composed of 4–5 individuals. If growth patterns
of *Litsea* and *Cascolaurus burmitis* are similar, then the latter would have been a
small to medium-sized tree.

Exalloanthum burmense, description based on Poinar 2018b, 2019a, b

Exalloanthum burmense is a very small delicate 2.1 mm long bisexual flower
with a 4.1 mm long pedicel (Fig. 1.55). The 5 slightly reclined tepals surround 10
elongate stamens whose filaments are connected at the base via a narrow ring.

Fig. 1.47 *Antiquifloria latifibris*

Fig. 1.48 Fleshy sepals of *Antiquifloria latifibris*

Fig. 1.49 Stamen of
Antiquifloria latifibris with
wide pointed connective

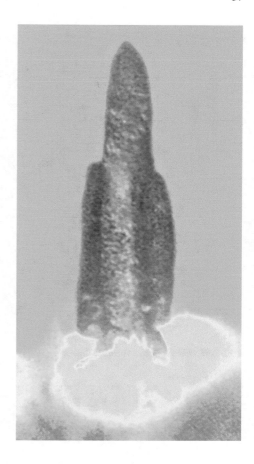

Fig. 1.50 Crane fly in Burmese amber

Fig. 1.51 *Cascolaurus burmitis* flower

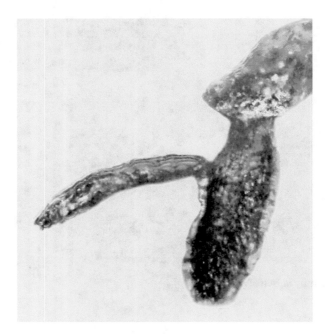

Fig. 1.52 Open anther flap of *Cascolaurus burmitis*

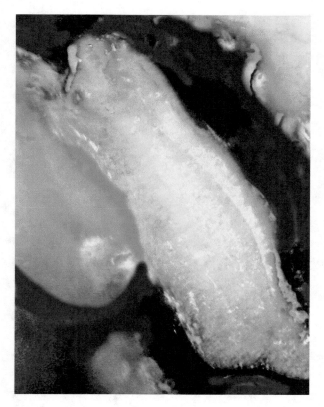

Fig. 1.53 *Cascolaurus burmitis* nectar glands

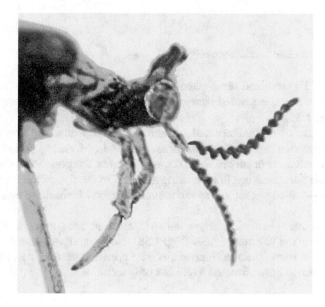

Fig. 1.54 Head of unicorn fly

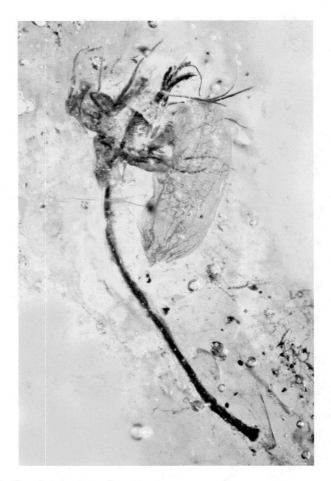

Fig. 1.55 *Exalloanthum burmense* flower

Adjacent is a large veined tepal possible representing a fallen petal. The pistil is composed of a large, expanded ovary with a short straight style capped by 5 stout stigmas (Fig. 1.56).

The structure of the anthers and method of pollen production and release are unique features of *Exalloanthum burmense*. Instead of forming pollen in anthers that open by valves as occurs in most flowering plants, the pollen forming tissue of *Exalloanthum burmense* is a thick coating attached to the tips of the filaments. The mature bullet-shaped pollen grains are released directly from these coated anthers (Fig. 1.57).

With the stamens extending above the style, an abundant supply of released pollen could fall on the viscous stigmas (Fig. 1.58). However, *Exalloanthum burmense* could also have been insect pollinated since an apparent nectar ring at the base of the stamens would have attracted a number of insects.

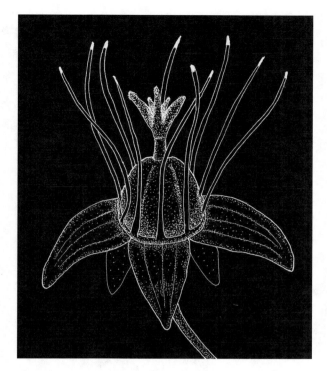

Fig. 1.56 Reconstruction of *Exalloanthum burmense*

While a few primitive members of the Magnoliales possess rod-shaped pollen grains, the manner of pollen formation and release is unique to *Exalloanthum burmense*, which cannot be assigned to any extant family of flowering plants.

Cyathitepala papillosa, description based on Poinar and Chambers 2020b

While *Cyathitepala papillosa* resembles the open wings of a butterfly, it is only because part of the flower was eaten before falling into the resin. Actually, the nibbling on flowers by insects may be how many blossoms detach and fall into fresh resin.

Like so many other blossoms in Burmese amber, *Cyathitepala papillosa* is a small bisexual flower with a width of just under 3 millimeters. Even though the flower is incomplete, enough of the crucial structures remain for a fairly complete description (Fig. 1.59). There are 6 large outer spirally arranged tepals that could be considered petals as well as numerous small inner tepals with densely papillate epidermis. A conic pistil bearing a pair of erect styles is surrounded by a central cluster of 12 stamens with papillose anthers that dehisce by 2 dorsally hinged valves (Fig. 1.60). The papillate surfaces of the sepals look like they have been sprinkled with small sand grains.

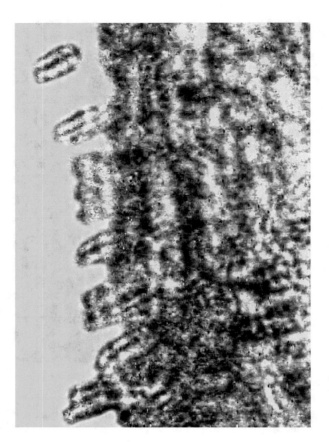

Fig. 1.57 Pollen release from anther of *Exalloanthum burmense*

While this flower has affinities with members of the Laurales, like so many flowers in Burmese amber, *Cyathitepala papillosa* cannot be placed in any existing family.

Chainandra zeugostylus, description based on Poinar and Chambers 2020a

With a width of 4.0 mm, almost twice as long as its height, the bisexual flower of *Chainandra zeugostylus* adds still another dimension to the assortment of Burmese amber flowers. There are 5 equal sepals, no petals and 10 stamens with sagittate bithecal anthers with short connectives and dorsal attaching filaments (Fig. 1.61). While most of its 10 stamens have had their roundish anthers removed, those remaining show that pollen was released by opened flaps on the anthers, with the thecae separate below the middle (Fig. 1.62). The style is single at the base, but two-branched above the middle with each branch bearing a terminal stigma (Fig. 1.63). The missing anthers were probably devoured by an insect herbivore since pollen is a high protein source. Surrounding the base of the ovary is a nectar disc that would have attracted pollinating insects. *Chainandra zeugostylus* also could not be placed in any present day family.

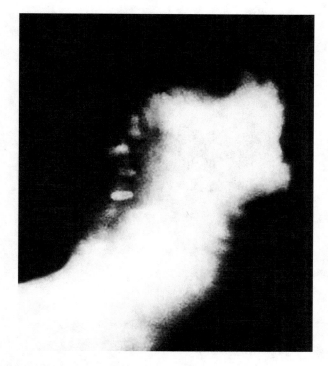

Fig. 1.58 Pollen attached to stigma of *Exalloanthum burmense*

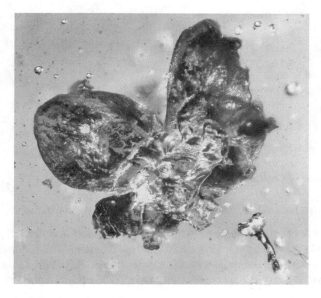

Fig. 1.59 Flower of *Cyathitepala papillosa*

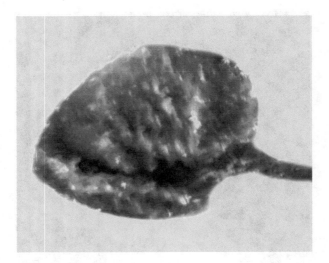

Fig. 1.60 Anther valves of *Cyathitepala papillosa*

Fig. 1.61 Flower of *Chainandra zeugostylus*

Strombothelya monostyla, description based on Poinar and Chambers 2019b

The flower of *Strombothelya monostyla* is 6 millimeters in width (Fig. 1.64). The smooth sepals have hairy margins and a strikingly reticulate venation. The 10 stamens are opposite and alternate with the sepals and strongly arched inwards around the single-styled ovary in the center of the flower. While the two portion of the ovary

Fig. 1.62   Stamen of *Chainandra zeugostylus* with dehising anther

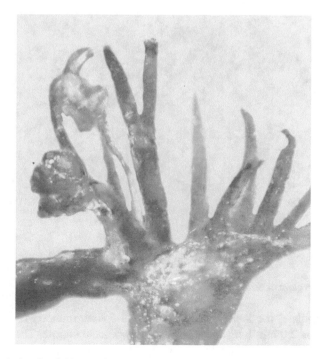

Fig. 1.63   Forked style of *Chainandra zeugostylus*

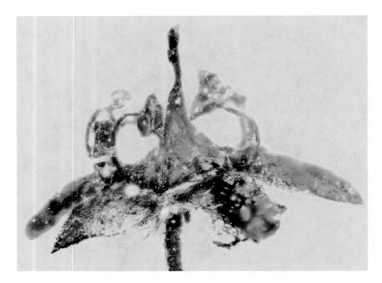

Fig. 1.64 Flower of *Strombothelya monostyla*

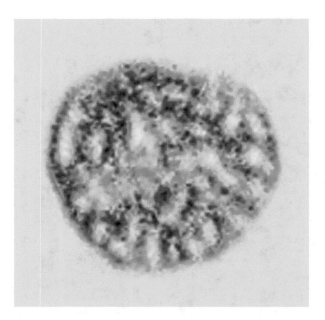

Fig. 1.65 Pollen grain of *Strombothelya monostyla*

is evenly papillate, the lower portion possesses 5 low, rounded ribs. The single straight style is 2.0 mm long and has a short- pubescent stigma. The pollen grains are spherical, 29 μm in diameter, with an alveolate exine and 3 indistinct colpi (Fig. 1.65).

Strombothelya grammogyna, description based on Poinar and Chambers 2019b

Strombothelya monostyla is considered closely related to *Strombothelya grammogyna* since both are bisexual flowers, have 5 spreading sepals but no petals, 10 arching stamens and a half-inferior ovary bearing either 1 (*Strombothelya monostyla*) or 3 (*Strombothelya grammogyna*) stout columnar styles capped with flat, hairy stigmas. The papillate surface of the top portion of the ovary of both species was probably nectariferous.

Strombothelya grammogyna is the smaller of the two, being only 4.0 millimeters in width. The surfaces and margins of the sepals are glabrous and have 3 distinct veins originating from the base. The 10 stamens are only slightly inclined around the ovary in the center of the flower (Fig. 1.66). The 3 diverging styles that form a loose triangle are 1.7 mm long and each possesses a pubescent stigma. The upper portion of the ovary is covered with minute papillae arranged in numerous lengthwise rows while the lower portion is 10-ribbed. The small papillae probably secreted nectar that attracted various insects. Possible pollinators could have been proboscis flies like *Cascomixticus tubuliferous* (Poinar and Vega 2021) (Fig. 1.67) that possessed tubular mouthparts for obtaining nectar and a small moth such as *Tanyglossus orectometopus* (Poinar 2017) (Fig. 1.68), whose descendants are known to seek nectar from flowers of the family Cunoniaceae, a group of shrubs, trees and

Fig. 1.66 Flower of *Strombothelya grammogyna*

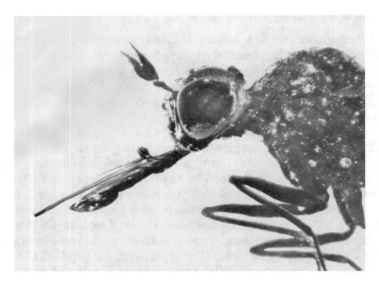

Fig. 1.67 Proboscis fly in Burmese amber

Fig. 1.68 Small moth in Burmese amber

stranglers that share some features with the 2 *Strombothelya* species. While there are some affinities with the Australasian family Cunionaceae, the two species of *Strombothelya* cannot be assigned to a modern family.

Phantophlebia dicycla, description based on Poinar and Chambers 2020d

With a width of 4.7 mm, *Phantophlebia dicycla* is one of the larger flowers in Burmese amber (Fig. 1.69). An interesting feature of *Phantophlebia dicycla* is that

Fig. 1.69   Flower of *Phantophlebia dicycla*

it possesses 10 distinctly vascularized tepals arranged in inner and outer whorls. They reveal an intricate venational pattern formed by 3 or 4 main veins interconnected by a network of veinlets. The 5 tepals of the inner whorl are wide, pointed and joined at the very base, being significantly larger than those of the outer (or lower) whorl and could be considered petals. While these inner tepals are brown in the fossil, when the flower was living, they could have been more vivid since hues are known to darken in amber. If colored, these tepals could have attracted solitary bees as well as other pollinators.

The 5 stamens, with their long slender filaments, bilocular anthers that dehise by lateral, oblique slits and papillate pollen sacs, are inserted opposite the inner tepals (Fig. 1.70). No gynoecium could be detected.

From a damaged anther and the presence of a beetle adjacent to one of the stamen filaments, we know that herbivores had already visited the flower. The beetle adjacent to the stamen could be a short-winged flower beetle whose adults and larvae are known to devour and/or develop on flowers today. Short-winged Flower beetles (Fig. 1.17) occur in Burmese amber and as their common name implies, the wings of these beetles are short and the tip of the abdomen is exposed. Their extended mandibles could easily reach the fragile anthers of *Phantophlebia dicycla*. This flower cannot be placed in any present-day plant family.

Eoëpigynia burmensis, description based on Poinar et al. 2007a

The small, bisexual regular flowers of *Eoëpigynia burmensis* are only 1.5 mm in length and have a series of short sepals followed by 4 petals with a reddish tinge (Fig. 1.71). The four stamens are in a single whorl, alternating with the petals. The filaments are linear and anthers introse and dorsifixed. The inferior ovary has a single stout style capped by a bilobed stigma (Fig. 1.72).

Fig. 1.70 Anther of *Phantophlebia dicycla*

The pollen grains are shed singly and possess three distinct paired thickened grooves (colpi) in the outer wall. With the bisexual condition, basal sepals, 4 separate petals alternating with the 4 stamens, an elongate receptacle and a single style with a bi-lobed stigma, the flower is provisionally assigned to the "dogwood" family Cornaceae *sensu lato*. Since the flower has distinctly tinted petals, it has an advantage competing for pollinating insects.

Eoëpigynia burmensis was in anthesis at the time it was entombed and pollen grains are on the anthers as well as the stigma. Some of the grains are so well preserved that they show both the pollination and fertilization nuclei. One of the petals of *Eoëpigynia burmensis* was infected with an actinobacterium with numerous groups of developing coccoidal cells scattered over the surface (Fig. 1.73). It is very rare to find identifiable pathogens on these early angiosperms, which brings up the question of how important are pathogenic fungi in determining the survival of various angiosperm lineages. We know that epiphyllous pycnidia infect leaves of angiosperms in Burmese amber (Poinar 2018d).

Endobeuthos paleosum, description based on Poinar and Chambers 2018a

Ranging from 3.0 to 4.2 mm in length, the 6 small, bisexual, tightly appressed flowers of *Endobeuthos paleosum* probably originated from the same inflorescence.

Fig. 1.71 Flower of
Eoëpigynia burmensis

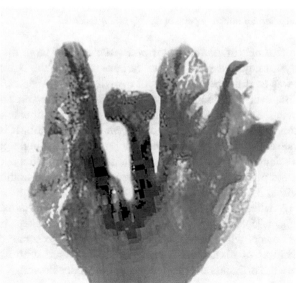

Fig. 1.72 Bilobed stigma of *Eoëpigynia burmensis*

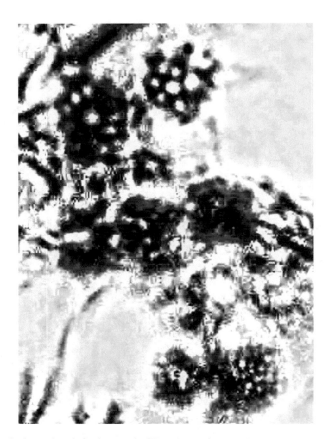

Fig. 1.73 Actinobacterium infecting petal of *Eoëpigynia burmensis*

Distinguishing features of *Endobeuthos paleosum* are the large number (50+) of small, helically arranged sepals and the variable sizes of the 5 imbricate petals (Fig. 1.74). It was as if the petals were still evolving in this species. In one flower, the petals are not even visible and in another flower only the spreading tips of the petals are showing. The other 3 flowers have their petals further extended and in the sixth flower, the petals are overlapping and form a floral sheath. It is a rare occurrence to have several flowers of the same inflorescence in a single piece of amber, but it provides scientists a view to the morphological variation that occurred in flowers from the same plant. The numerous stamens have enlarged anther connectives. Ovaries with either 3 or 4 styles are present.

An immature orthopteran is preserved adjacent to one of the *Endobeuthos paleosum* flowers (Fig. 1.75). Like present day grasshoppers and crickets, it was probably feeding on the flowers when the short inflorescence fell into some resin. Since the vegetative structures of *Endobeuthos paleosum* are covered with hard, spiny surfaces, it is likely the orthopteran was devouring the tender flowers.

Fig. 1.74 Two flowers of
Endobeuthos paleosum

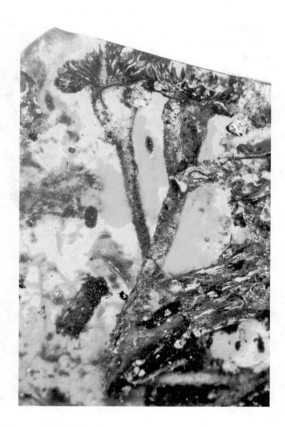

While flowers of *Endobeuthos paleosum* differ from those of any known modern day family, the numerous sepals and stamens, structure of the anthers and presence of petals affiliate the genus with the Australian fire vine trees (Dillenaceae).

Chenocybus allodapus, description based on Poinar 2018b

The "flower" of *Chenocybus allodapus*, with a length of 7.0 mm and width of 7.2 mm, is quite unique and its structure is not patterned after that of any other angiosperm flower in or out of Burmese amber. The "flower" could also be considered an inflorescence with the ovaries and stamens borne on separate pedicels. There are three short, connate, bract-like "sepals" located at the base of the peduncle and 12 elongated, thick, spirally arranged, unequal "petals" (Figs. 1.76 and 1.77).

The stamens and pistils are equally strange since they are located at the apex of extended narrow stalks arising from the petal clusters (Fig. 1.76). The anthers are shield-like, open by longitudinal slits and contain spherical pollen. The superior ovaries lack a style but have a stigmatic crest. Some of the ovaries are mature with dehisced fruitlets containing seeds (Fig. 1.78). It is not possible to align *Chenocybus allodapus* with any known plant family.

Fig. 1.75 *Endobeuthos paleosum* flower with insect

While the walls of the pistils and stamens appear to be quite thick, some herbivores were still able to break through these barriers to reach the pollen and developing seeds. The herbivores could have been large beetles that also served as pollinators simply by their random movements while devouring portions of the flowers.

Such damage shows how herbivorous insects on one hand, and beneficial pollinators on the other hand, can control the fate of evolving angiosperms. While establishing features to attract dependable pollinators and seed dispersers, early angiosperms also had to develop structural, chemical and developmental features for protection against various herbivores.

Mirafloris burmitis, description based on Poinar 2021b

Mirafloris burmitis represents the first showy monocot flower in Burmese amber (Fig. 1.79). The perianth consists of two whorls of 3 + 3 separate tepals, two whorls of 3 + 3 free stamens positioned opposite the tepals and a single 3-lobed style in the center of the flower. Other monocot characters are the perigonal nectaries located at the base of the inner tepals and the monocolpate pollen (Fig. 1.80). These features, along with the position of the stamens in relation to the style, are commonly found in members of the family Liliaceae and the fossil is tentatively assigned to that family.

Fig. 1.76 Complete flower of *Chenocybus allodapus* showing sepals (arrow), petals (black structures), stamens (white structures) and pistils (grey structures)

A curious feature of *Mirafloris burmitis* is that the median tepal is larger than the remainder, thus making the flower weakly bisymmetric or zygomorphic. While pollinators are unknown, small rove beetles (Staphylinidae: Coleoptera) had established themselves in the flower before it entered the resin and one was found at the base of a stamen and another inside an anther. Another anther that had been partially eaten was adjacent to the flower, revealing rows of pollen.

The appearance of the family Liliaceae was estimated to be around 90 mya, which would make *Mirafloris burmitis* one of the earliest lineages of the family. The Liliaceae is thought to have originated in South Gondwana, which lends support to the theory that fossils in Burmese amber had a Gondwanan origin.

Eophylica priscastellata, description based on Shi et al. 2022, Poinar and Chambers (in press), and ***Phylica piloburmensis***, description based on Shi et al. 2022

Members of the genera *Phylica* Linn. and *Eophylica* Shi, Wang & Engel are unusual members of the family Rhamnaceae. Mature plants have terminal flowers

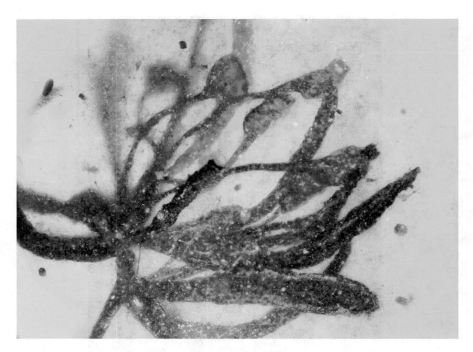

Fig. 1.77 Petals, stamens and pistils of *Chenocybus allodapus*

closely surrounded by an involucre of erect, acuminate bracts (Fig. 1.81). Members of the genus *Phylica* posess 5 sepals and 5 petals while those of *Eophylica* possess 8 sepals and lack petals (Shi et al. 2022).

The accompanying elongate, slender leaves are covered with densely stellate-pubescent appendages or spines (Fig. 1.82). Such stellate clusters, both stalked and sessile, which are also abundant on the bracts and sepals, are a diagnostic character of both genera. They are reminescent of miniature cactus spines and possibly served to protect the plant from herbivores.

Without visible flowers, the true identity of these plants is well hidden. The present author erroneously described the vegetatively stages of one of these plants as a terrestrial alga (Poinar and Brown 2019). Another specimen with a flower was considered to represent the seed of a member of the gymnosperm family Gnetaceae (Xia et al. 2015).

Several specimens of *Eophylica priscastellata* from the present authors collection were further characterized (Poinar and Chambers in press). While petals are absent, the flower heads, which range from 1.0 to 3.0 mm in length, are closely surrounded by an involucre of erect, acuminate bracts. The 6 exposed stamens have elongate, linear-oblong, bilocular anthers with a stout pointed connective (Fig. 1.83). The filaments are short and wide. The inferior ovary is globose and the stigma is often hidden from view behind extraneous stellate trichomes. A tissue layer on the top of the ovary is probably an epigynous disc. The describers of the genus (Shi

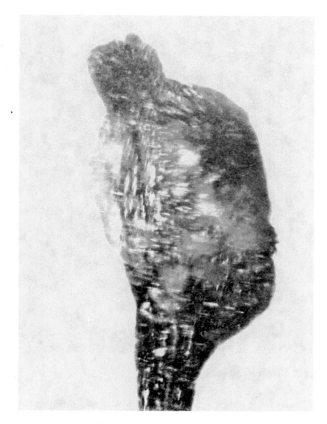

Fig. 1.78 Mature ovary of *Chenocybus allodapus*

et al. 2022) associate the fossils with the history of fire-adapted angiosperms in Cretaceous Gondwana.

In addition to the wide range of flowers depicted above are two additional flowers in Burmese amber. The first, *Jamesrosea burmensis,* is a small bisexual flower (about 2.1 mm in diameter) with a hemispheric floral cup bearing some 12 imbricate ovate sepals, no petals, but 12 stamens arranged in a tight spiral (Crepet et al. 2016). This flower possesses both fertile stamens with long filaments and shorter sterile stamens (staminodes). The ovary is 4 parted with 4 tapering styles capped with slightly expanded stigmas. The authors concluded that the flower was a member of the Laurales and falls within the Siparunaceae-Athersopermataceae clade.

The second flower, *Lijinganthus revolute*, is bisexual with 5 small sepals and 5 large recurved petals (Liu et al. 2018). Here is a flower that has already formed well-developed petals to attract pollinators. The 8–10 stamens have long, slender filaments and the ovary is three-parted. This is one of the larger flowers reported from Burmese amber, with a length of 6.5 mm and a width of 4.8 mm. The authors did not suggest the relationship of *Lijinganthus revolute* with any known plant family.

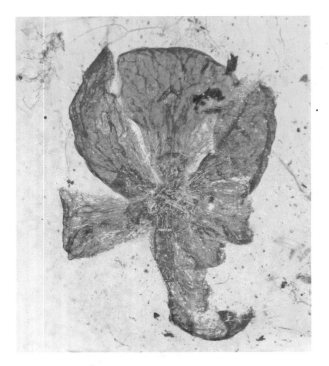

Fig. 1.79 Flower of *Mirafloris burmitis*

Fig. 1.80 Monocolpate
pollen of *Mirafloris
burmitis*

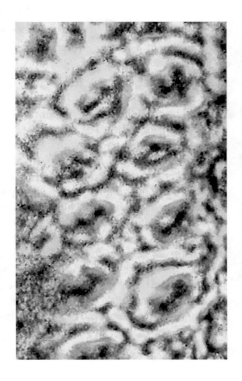

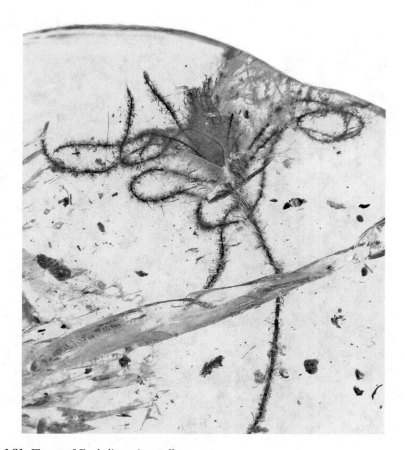

Fig. 1.81 Flower of *Eophylica priscastellata*

Fig. 1.82 Leaf of Eophylica priscastellata covered with stellate spines

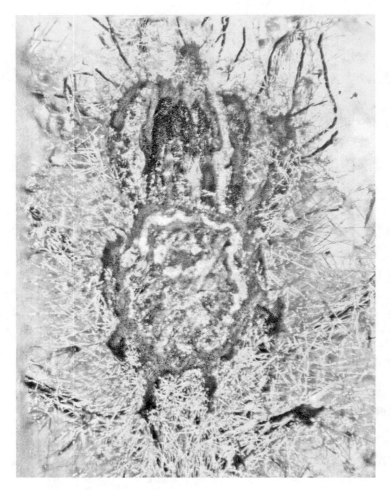

Fig. 1.83 Flower head of *Eophylica priscastellata*

1.2   Summary of Burmese Amber Flowers

The above diverse assortment of flowers that were present in the Burmese amber forest is just a fraction of the angiosperms that actually existed at that time. But they give us a glimpse of some of the variability of flowering plants that flourished during the reign of the dinosaurs and show how they differ from most of today's flora (Table 1.1).

Now we know that a range of both monocots and dicots displaying perfect and imperfect flowers were well established by the mid-Cretaceous. Many lacked definite petals, but all had variously arranged sepals (or tepals), some of which

Table 1.1 Burmese amber flowers (N = 31)

Flower	Systematic placement	References
Antiquifloria latifibris	Unknown	Poinar et al. (2016)
Cascolaurus burmitis	Lauraceae	Poinar (2016)
Chainandra zeugostylus	Unknown	Poinar and Chambers (2020a)
Chenocybus allodapus	Unknown	Poinar (2018b)
Cyathitepala papillosa	Laurales	Poinar and Chambers (2020b)
Dasykothon leptomiscus	Atherospermataceae	Poinar and Chambers (2020c)
Dispariflora robertae	Laurales	Poinar and Chambers (2019c)
Endobeuthos paleosum	Dillenaceae	Poinar and Chambers (2018a)
Eoëpigynia burmensis	Cornaceae	Poinar et al. (2007a)
Eophylica priscastellata	Rhamnaceae	Shi et al. (2022) and Poinar and Chambers (2022)
Exalloanthum burmense	Unknown	Poinar (2018c, 2019a, b)
Jamesrosea burmensis	Laurales	Crepet et al. (2016)
Lachnociona camptostylus	Brunelliaceae, Cunoniaceae	Poinar and Chambers (2018c)
Lachnociona terriae	Brunelliaceae, Cunoniaceae	Poinar et al. (2008)
Lijinganthus revolute	Unknown	Liu et al. (2018)
Micropetasos burmensis	Unknown	Poinar et al. (2013)
Mirafloris burmitis	Liliales	Poinar (2021b)
Palaeoanthella huangii	Monimiaceae	Poinar and Chambers (2005)
Phantophlebia dicycla	Unknown	Poinar and Chambers (2020d)
Phylica piloburmensis	Rhamnaceae	Shi et al. (2022)
Programinis burmitis	Poaceae	Poinar (2004) and Poinar et al. (2015)
Programinis laminatus	Poaceae	Poinar (2004, 2011)
Setitheca lativalva	Laurales	Poinar and Chambers (2018b)
Strombothelya grammogyna	Unknown	Poinar and Chambers (2019b)
Strombotheyla monostyla	Unknown	Poinar and Chambers (2019b)
Thymolepis toxandra	Monimiaceae	Chambers and Poinar (2020)
Tropidogyne euthystyla	Cunionaceae	Poinar et al. (2021)
Tropidogyne lobodisca	Cunionaceae	Poinar and Chambers (2019a)
Tropidogyne pentaptera	Cunionaceae	Poinar and Chambers (2017)
Tropidogyne pikei	Cunionaceae	Chambers et al. (2010)
Valviloculus pleristaminis	Monimiaceae	Poinar et al. (2020)
Zygadelphus aetheus	Hernandiaceae	Poinar and Chambers (2019d)

apparently played the role of petals in attracting pollinators. Just as today's flowers, those in Burmese amber possessed few to many stamens as well as staminodes, and the variously shaped anthers exhibited different methods of pollen dehiscence. The pistils were also quite variable with inferior and superior ovaries and curved, erect, single or branched styles with terminal or sub-terminal stigmas.

Based on their variety, it is obvious that the mid-Cretaceous was a period of floral experimentation with only a few of the lineages producing descendants that are still surviving today, such as the grasses (Poaceae) and laurels (Lauraceae). At this time angiosperms were "experimenting" with different ways to attract pollinators as part of their adaptation to varying habitats. Many Burmese amber flowers had some type of nectar producing glands while others attracted pollinators by odors and oil production (Poinar and Poinar 2018).

We will never know how long the various floral lineages shown here survived in the Burmese amber forest nor what features some lineages developed that resulted in their survival today. Strong affinities of many Burmese flowers (representatives of the Monimiaceae, Cunionaceae and Dillenaceae) align them with ancient families in relictual Austrlian rain forests and in previous Gondwana landmasses (Poinar 2018a). White (1986, 1994) discusses reasons for believing that western Gondwana (Australia) was the birthplace of angiosperms some 120 mya. It would be interesting to know what adaptive factors in the genetic makeup of these lineages allowed them to survive to the present. The causes of angiosperm extinctions are manyfold (e.g. herbivory, disappearance of pollinators, changing climate, diseases, habitat deterioration, competition with other plants, absence of compatable mycorrhiza, etc.).

While many structural features of Burmese amber flowers are similar to those of today's blooms, some developmental characters have not been noted in extant angiosperms. For example, there are no flowers today that have sepals of such diverse shapes and sizes and as *Dispariflora robertae* and no blooms are known that produce secondary anthers from the backs of primary anthers, as in *Zygadelphus aetheus*.

Key to Burmese Amber Flowers

1. Leaves grass-like, flowers not present – *Programinis laminatus*

    1A. Flower present – 2

2. Flowers in spikelets – *Programinis burmitis*

    2A. Flowers not in spikelets – 3

3. Petals absent – 4

    3A. With petals or outer large petal-like tepals – 10

4. Flowers with 30+ short, upward-pointing stamens – *Valviloculus pleristaminis*

    4A. Flowers not as above – 5

5. Flowers with different sized sepals; new flowers developing from branches of old flowers – *Dispariflora robertae*

    5A. Flowers not as above – 6

6. Flowers with 8 sepals that have the outer surfaces bearing stalked and sessile, stellate spines – *Eophylica priscastellata*

   6A. Flowers with 5 spreading, reticulately-veined sepals, inferior 10-veined ovary – 7

7. Flowers with 10 stamens and 3 short, curved styles – *Tropidogyne pikei*

   7A. Flowers not as above – 8

8. Flowers with 5 stamens and 2 slender styles – *Tropidogyne pentaptera*

   8A. Flowers not as above – 9

9. Flower pistillate, sepals glabrous – *Tropidogyne lobodisca*

   9A. Flower bisexual, sepals puberulent – *Tropidogyne euthystyla*

10. Flowers with 5 large sepals and 2 backward diverging separate styles – 11

    10A. Flowers not as above – 12

11. Flower bisexual with 4 styles – *Lachnociona terriae*

    11A. Flower pistillate with 5 styles – *Lachnociona camptostylus*

12. With 5 imbricate petals and 50+ helically arranged sepals – *Endobeuthos paleosum*

    12A. Flowers not as above – 13

13. Flowers with 12 elongated, thick, spirally arranged, unequal petals – *Chenocybus allodapus*

    13A. Flowers not as above – 14

14. Flowers with 5 small sepals and 5 large recurved petals – 15

    14A. Flowers not as above – 16

15. Sepals with stellate spines; anthers enfolded by petals – *Phylica piloburmensis*

    15A. Sepals withouy stellate spines, anthers not enfolded by petals – *Lijinganthus revolute*

16. Flowers with a series of short sepals and 4 petals – *Eoëpigynia burmensis*

    16A. Flowers not as above – 17

17. Flower staminate, with 12 spirally arranged tepals and 10 mushroom-shaped stamens – *Setitheca lativalva*

    17A. Flowers not as above – 18

18. Flower staminate, with ring of 8 subsessile stamens alternating with 8 sepals – *Palaeoanthella huangii*

    18A. Flowers not as above – 19

19. Flower bisexual with 12 variously sized tepals and 2 stamens – *Thymolepis toxandra*

    19A. Flowers not as above – 20

20. Flowers bisexual, with 5 irregular sepals, hooked styles and up to 60 stamens – *Micropetasos burmensis*

    20A. Flowers not as above – 21

21. Flower bisexual, with 19 extended spirally arranged fleshy sepals and 12 stamens – *Antiquifloria latifibris*

    21A. Flowers not as above – 22

22. Flower staminate with 6 tepals in 2 sets of three; 9 stamens in 3 whorls; anthers with valve flaps – *Cascolaurus burmitis*

    22A. Flowers not as above – 23

23. Flower bisexual with 5 tepals, 10 stamens in 2 whorls; style with 5 stigmatic branches; elongate pollen grains released through anther walls – *Exalloanthum burmense*

    23A. Flowers not as above – 24

24. Flower bisexual with 6 large outer petal-like tepals and numerous small inner tepals; 12 stamens dehising by dorsally hinged valves – *Cyathitepala papillosa*

    24A. Flowers not as above – 25

25. Flower with 5 equal tepals, 10 stamens with sagittate bithecal anthers opening by flaps; style 2 branched – *Chainandra zeugostylus*

    25A. Flowers not as above – 26

26. Flowers bisexual with 5 spreading sepals, 10 arching stamens and half inferior ovary – 27

    26A. Flowers not as above – 28

27. Sepals with reticulate venation and hairy margins; single straight style – *Strombotheyla monostyla*

    27A. Flowers with sepals with 3 distinct veins; with 3 diverging styles – *Strombotheyla grammogyna*

28. Flower with 10 distinctly vascularized tepals in 2 whorls; 5 stamens with long, slender filaments – *Phantophlebia dicycla*

    28A. Flowers not as above – 29

29. Flower bisexual, with 3 oval and 2 narrow sepals, 12 stamens and 2 elongate styles – *Dasykothon leptomiscus*

    29A. Flowers not as above – 30

30. Flower bisexual with 10 spirally arranged tepals, 2 styles and secondary anthers formed on back of primary anthers – *Zygadelphus aetheus*

    30A. Flowers not as above – 31

31. Flower with 2 whorls of 3 + 3 tepals, 3 + 3 free stamens, 3-lobed style and perigonal nectaries – *Mirafloris burmitis*

    31A. Flower with 12 imbricate ovate sepals and 12 stamens arranged in a tight spiral; 4 parted ovary with 4 tapering styles – *Jamesrosea burmensis*

References

Chambers KL, Poinar GO Jr (2020) *Thymolepis toxandra* gen. et sp. nov., a mid-Cretaceous fossil flower with horseshoe-shaped anthers. J Bot Res Inst Texas 14:57–64

Chambers KL, Poinar GO Jr, Buckley RT (2010) *Tropidogyne*, a new genus of Early Cretaceous eudicots (Angiospermae) from Burmese amber. Novon 20:23–29

Crepet WL, Nixon KC, Grimaldi D, Ricco M (2016) A mosaic Lauralean flower from the Early Cretaceous of Myanmar. Am J Bot 103:290–297

Cruickshank RD, Ko K (2003) Geology of an amber locality in the Hukawng Valley, northern Myanmar. J Asian Earth Sci 21:441–455

Danforth BD, Poinar GO Jr (2011) Morphology, classification, and antiquity of *Melittosphex burmensis* (Apoiodea: Melittosphecidae) and implications for early bee evolution. J Paleontol 85:882–891

Liu Z-J, Huang D, Cai C, Wang X (2018) The core eudicot boom registred in Myanmar amber. Nat Sci Rep. https://doi.org/10.1038/s41598-018-35100-4

Poinar GO Jr (2004) *Programinis burmitis* gen. et sp. nov., and *P. laminatus* sp. nov., Early Cretaceous grass-like monocots in Burmese amber. Aust Syst Bot 17:497–504

Poinar GO Jr (2010) *Cascoplecia insolitis* (Diptera: Cascopleciidae), a new family, genus and species of flower-visiting, unicorn fly (Bibionomorpha) in Early Cretaceous Burmese amber. Cretac Res 31:71–76

Poinar GO Jr (2011) Silica bodies in the Early Cretaceous *Programinis laminatus* (Angiospermae: Poales). Palaeodiversity 4:1–6

Poinar GO Jr (2016) A mid-Cretaceous Lauraceae flower, *Cascolaurus burmitis* gen. et sp. nov., in Myanmar amber. Cretac Res 71:96–101

Poinar GO Jr (2017) A new genus of moths (Lepidoptera: Gracillarioidea: Douglasiidae) in Myanmar amber. Hist Biol 31:1–5

Poinar GO Jr (2018a) Burmese amber: evidence of Gondwanan origin and Cretaceous dispersion. Hist Biol 31:1304–1309

Poinar GO Jr (2018b) Mid-Cretaceous Angiosperm flowers in Myanmar amber (*Chenocybus allo-dapus*). In: Welch B, Wilkerson M (eds) Recent advances in plant research. Nova Science Publishers, New York, pp 190–195. ISBN: 978-1-53614-170-2

Poinar GO Jr (2018c) Mid-Cretaceous Angiosperm flowers in Myanmar amber (*Exalloanthum* (*Diaphoranthus*) *burmense*). In: Welch B, Wilkerson M (eds) Recent advances in plant research. Nova Science Publishers, New York, pp 196–204. ISBN: 978-1-53614-170-2

Poinar GO Jr (2018d) A mid-Cretaceous pycnidia, *Palaeomycus epallelus* gen. et sp. nov., in Myanmar amber. Hist Biol 32:234–237

Poinar GO Jr (2019a) *Exalloanthum*, a new name for a fossil angiosperm (*Diaphoranthus*) in Myanmar amber. J Bot Res Inst Texas 13:475–476

Poinar GO Jr (2019b) A new genus of wasp (Hymenoptera: Evanioidea: Praeaulacidae) associated with an angiosperm flower in Burmese amber. Palaeoentomology 2:474–481

Poinar GO Jr (2020) Discoscapidae fam. nov. (Hymenoptera: Apoidea), a new family of stem lineage bees with associated beetle triungulins in mid-Cretaceous Burmese amber. Palaeodiversity 12:1–9

Poinar GO Jr (2021a) A new fern, *Cladarastega burmanica* gen. et sp. nov. (Dennstaedtiaceae: Polypodiales) in mid-Cretaceous Burmese amber. Palaeodiversity 14:153–160

Poinar GO Jr (2021b) A monocot flower, *Mirafloris burmitis* gen. et sp. nov. in Burmese amber. Biosis: Biol Syst 2:1–7

Poinar GO Jr, Brown AE (2017) A new genus of leafhoppers (Hemiptera: Cicadellidae) in mid-Cretaceous Myanmar amber. Hist Biol 32:160–163

Poinar GO Jr, Brown AE (2018) *Furcalabratum burmanicum* gen. et sp. nov. a short-winged flower beetle (Coleoptera: Kateretidae) in mid-Cretaceous Myanmar amber. Cretac Res 84:1–5

Poinar GO Jr, Brown AE (2019) A green algae (Chaetophorales: Chaetophoraceae) in Burmese amber. Hist Biol 33:1–5

Poinar GO Jr, Chambers KL (2005) *Palaeoanthella huangii* gen. and sp. nov., an early Cretaceous flower (Angiospermae) in Burmese amber. Sida 21:2087–2092

Poinar GO Jr, Chambers KL (2017) *Tropidogyne pentaptera*, sp. nov., a new mid-Cretaceous fossil angiosperm flower in Burmese amber. Paleodiversity 10:135–140

Poinar GO Jr, Chambers KL (2018a) *Endobeuthos paleosum* gen. et sp. nov., fossil flowers of uncertain affinity from mid-Cretaceous Myanmar amber. J Bot Res Inst Texas 12:133–139

Poinar GO Jr, Chambers KL (2018b) *Setitheca lativalva* gen. et sp. nov., a fossil flower of Laurales from mid-Cretaceous Myanmar amber. J Bot Res Inst Texas 12:643–653

Poinar GO Jr, Chambers KL (2018c) Fossil flowers of *Lachnociona camptostylus* sp. nov., a second record for the genus in mid-Cretaceous Myanmar amber. J Bot Res Inst Texas 12:655–666

Poinar GO Jr, Chambers KL (2019a) *Tropidogyne lobodisca* sp. nov., a third species of he genus from mid-Cretaceous Myanmar amber. J Bot Res Inst Texas 13:461–466

Poinar GO Jr, Chambers KL (2019b) *Strombothelya* gen. nov., a fossil angiosperm with two species in mid-Cretaceous Myanmar amber. J Bot Res Inst Texas 13:451–460

Poinar GO Jr, Chambers KL (2019c) *Dispariflora robertae* gen. et sp. nov., a mid-Cretaceous flower of possible Lauralean affinity from Myanmar amber. J Bot Res Inst Texas 13:173–183

Poinar GO Jr, Chambers KL (2019d) *Zygadelphus aetheus* gen. et sp. nov., an unusual fossil flower from mid-Cretaceous Myanmar amber. J Bot Res Inst Texas 13:467–473

Poinar GO Jr, Chambers KL (2020a) *Chainandra zeugostylus* gen. et sp. nov., a mid-Cretaceous amber fossil with a novel mode of valvate anther dehiscence. J Bot Res Inst Texas 14:367–372

Poinar GO Jr, Chambers KL (2020b) *Cyathitepala papillosa* gen. et sp. nov., a mid-Cretaceous fossil flower from Myanmar amber with valvate anthers. J Bot Res Inst Texas 14:351–358

Poinar GO Jr, Chambers KL (2020c) *Dasykothon leptomiscus* gen. et sp. nov., a fossil flower of possible Lauralean affinity from Myanmar amber. J Bot Res Inst Texas 14:65–71

Poinar GO Jr, Chambers KL (2020d) *Phantophlebia dicycla* gen. et sp. nov., a five-merous fossil flower in mid-Cretaceous Myanmar amber. J Bot Res Inst Texas 14:73–80

Poinar GO Jr, Chambers KL (2022) Additional fossil specimens of *Eophylica* (Rhamnaceae) in mid-Cretaceous amber from the Hukawng valley, Myanmar. J Bot Res Inst Texas (in press)

Poinar GO Jr, Danforth BN (2006) A fossil bee from Early Cretaceous Burmese amber. Science 314:614

Poinar GO Jr, Poinar GR (2018) The antiquity of floral secretory tissues that providetoday's fragrances. Hist Biol 32:1–6

Poinar GO Jr, Vega FE (2020) A new genus of cylindrical bark beetle (Coleoptera: Zopheridae: Colydiinae) in mid-Cretaceous Burmese amber. Biosis: Biol Syst 1:134–140

Poinar GO Jr, Vega FE (2021) A new genus of Apsilocephalidae (Diptera) in mid-Cretaceous Burmese amber. Biosis: Biol Syst 2:1–7

Poinar GO Jr, Chambers KL, Buckley R (2007a) *Eoëpigynia burmensis* gen. and sp. nov., an Early Cretaceous eudicot flower (Angiospermae) in Burmese amber. J Bot Res Inst Texas 1:91–96

Poinar GO Jr, Lambert JB, Wu Y (2007b) Araucarian source of fossiliferous Burmeses amber: spectroscopic and anatomical evidence. J Bot Res Inst Texas 1:449–455

Poinar GO Jr, Gorochov AV, Buckley R (2007c) Longioculus burmensis, n. gen., n. s (Orthoptera: Elcanidae) in Burmese amber. Proc Entomol Soc Wash 109:649–655

Poinar GO Jr, Chambers KL, Buckley RT (2008) An Early Cretaceous angiosperm fossil of possible significance in rosid floral diversification. J Bot Res Inst Texas 2:1183–1191

Poinar GO Jr, Chambers KL, Wunderlich J (2013) *Micropetasos*, a new genus of Angiosperms from mid-Cretaceous Burmese amber. J Bot Res Inst Texas 7:745–750

Poinar GO Jr, Alderman S, Wunderlich J (2015) One hundred million year old ergot: psychotropic compounds in the Cretaceous? Palaeodiversity 8:13–19

Poinar GO Jr, Buckley R, Chen H (2016) A primitive mid-Cretaceous angiosperm flower, *Antiquifloris latifibris* gen. & sp. nov., in Myanmar amber. J Bot Res Inst Texas 10:55–162

Poinar GO Jr, Chambers KL, Vega FE (2020) *Valviloculus pleristaminis* gen. et sp. nov., a mid-Cretaceous amber flower related to families in Order Laurales. J Bot Res Inst Texas 14:359–366

Poinar GO Jr, Chambers KL, Vega FE (2021) *Tropidogyne euthystyla* sp. nov., a new small-lowered addition to the genus from mid-Cretaceous Myanmar amber. J Bot Res Inst Texas 15:113–119

Shi G, Grimaldi DA, Harlow GE, Wang J, Wang J, Yang M, Lei W, Li Q, Li X (2012) Age constraint on Burmese amber based on U-Pb dating of zircons. Cretac Res 37:155–163

Shi C et al (2022) Fire-prone Rhamnaceae with South African affinities in Cretaceous Myanmar amber. Nat Plant. https://doi.org/10.1038/s41477-021-01091-w

White ME (1986) The greening of Gondwana. Reed Books, Frenchs Forest, 356 pp

White ME (1994) After the greening; the browning of Australia. Kangaroo Press, Kenthurst, 288 pp

Xia FY, Yang G, Zhang Q, Shi GL (2015) Amber time lines through time and space. Beijing Science Press. (In Chinese)

Chapter 2
Baltic Amber Flowers

Abstract Historically, Baltic amber is the most famous of all the amber deposits in the world. Based on leaves, flowers and seeds, some 100 angiosperms have been described from this amber source. After discussing some of the descriptions from the nineteenth and twentieth centuries, this work shows recently described angiosperm flowers in Baltic amber. One of these is the first grass described from these deposits. Intracellular remains in the spikelet were so well preserved they could be compared with present day members. Also in Baltic amber is the oldest known orchid, represented by coherent masses of pollen grains (pollinarium) attached to a fungus gnat. Based on the remains of leaves, flowers and seeds of both angiosperms and gymnosperms preserved in Baltic amber, a scenario of the original amber forest some 35–40 mya is presented.

Studies on Baltic amber began with Pliny the Elder in the first century AD. Named succinite in his book on Natural History, it remains the most studied of all amber deposits, desired not just for fossils but originally for carvings, beads and a wide assortment of jewelry.

It is still possible to find pieces of Baltic amber today washed up along the shoreline of the North Sea (Fig. 2.1) or on the beaches of the Baltic Sea. This is especially prevalent after storms, when wave action loosens pieces of amber from the exposed "blue earth" layer located some 15–20 feet below sea level along the Baltic Sea. Presently, mining operations located along the Baltic Sea coast, especially in the vicinity of the Samland Peninsula and at the Yantarny Amber Quarry near Kaliningrad provide most fossiliferous Baltic amber.

A large number of plant and animal fossils have been described from Baltic amber, including no less than 100 angiosperms (Caspary and Klebs 1907; Czeczott 1961; Conwentz 1886, 1890; Larsson 1978; Weitschat and Wichard 2002). The location of many of these specimens, if they still exist, is unknown. A few private collections of Baltic amber fossils have been recovered, showing storage methods at that time (Fig. 2.2). For the systematic placement of these previously described flowers, we refer the reader to the list provided by Czeczott (1961). The present work primarily deals with recently described angiosperms in Baltic amber.

G. Poinar, *Flowers in Amber*, Fascinating Life Sciences,
https://doi.org/10.1007/978-3-031-09044-8_2

Fig. 2.1 In Jutland, Denmark, Baltic amber occurs along the shore of the North Sea

Based on fossils collected from numerous locations throughout northern Europe, the Baltic amber forest extended from Scandinavia through Eastern Russia to Germany. And based on the environment of present day descendants of Baltic amber organisms, the Baltic amber forest experienced both a warm temperate climate and a tropical or subtropical climate. Some scientists explain the disparity of life forms in Baltic amber, with both temperate and tropical descendants, to a global climate change that occurred during the existence of the amber forest, which may have been millions of years (Wolfe et al. 2016). Others suggest that the forest may have extended from sea level to the alpine zone. Either theory could explain the wide geographical locations of many present day descendants of Baltic amber organisms. Amber found near Rovno (Rivne), a city in western Ukraine, is considered coeval to that of Baltic amber, and probably represents a southern extension of the Baltic amber forest. Since there are few common species in Baltic and Rovno amber, each region was probably subjected to different climatic conditions. A single male flower assigned to the genus *Prunus* L. has been described from Rovno amber, indicating the presence of warm winters at that period (Sokoloff et al. 2018).

The widespread remains of pine needles and oak hairs in Baltic amber, showing that pines and members of the oak family (Fagaceae) were fairly common, indicate a warm temperate climate. The evidence of spore-bearing plants, such as ancient cypresses, ferns, cycads, podocarps, yews, spruces, firs, larch, junipers and red-woods, as well as the presence of swamps and marshes indicate a warm temperate climate of the Baltic amber forest. What have been identified as maples, magnolias, roses, willows, nettles, legumes, euphorbs, campanulas, laurals, lilies and moor

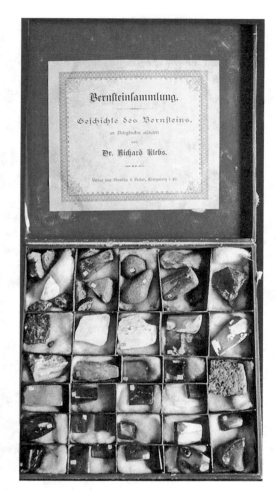

Fig. 2.2 Private Baltic amber collection of Richard Klebs (1850–1911)

grasses also suggest a warm temperate climate. The presence of a few plants, like olives, palms and kauri trees (*Agathis*) suggest a subtropical-tropical habitat (Czeczott 1961; Conwentz 1886; Larsson 1978; Weitschat and Wichard 2002).

Chemical tests on Baltic amber indicate that kauri trees were responsible for the majority, if not all, of the amber in the Baltic amber forest and possibly the Rovno amber forest. This is not saying that kauri trees were present during the entire existence of the amber forest or that other conifers did not also contribute to amber formation since the time period of amber formation in the Baltic amber forest is unknown. Kauri trees could have perished long before the forest disappeared or became Boreal. Some insects in Baltic amber, such as a small weevil (*Oxycraspedus poinari* Legalov 2016) (Fig. 2.3) whose descendants feed on Araucariaceae today, support the presence of kauri trees in the Baltic amber forest.

Fig. 2.3 Kauri weevil in Baltic amber

Fig. 2.4 Baltic amber fungus gnat

To determine how well Baltic amber preserves internal tissues of captured insects, a fungus gnat (Mycetophilidae) (Fig. 2.4) was examined under the electron microscope. Surprisingly, muscle fibers, ribosomes, lipid droplets, mitochrondia and even nuclei with chromatin were still present (Fig. 2.5). In fact, these ancient cellular structures appeared very much in the same condition as in recent insects. This shows the amazing preservation properties of amber (Poinar and Hess 1982).

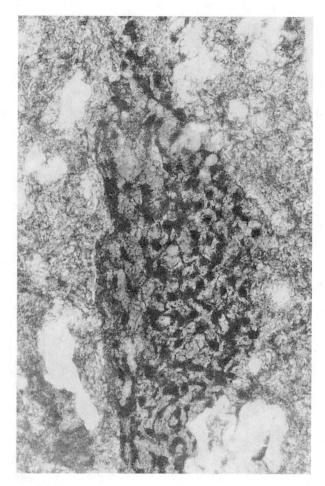

Fig. 2.5 Nucleus in Baltic amber fungus gnat

2.1 Flora in Baltic Amber

2.1.1 *Fagaceae*

The "oak" hairs (Fig. 2.6) commonly found in Baltic amber could well have come from species of *Quercus*, but some may be from other genera of Fagaceae such as *Fagus* or *Castanea*. The same applies to both male (Fig. 2.7) and female (Fig. 2.8) "oak" flowers.

Oaks of the genus *Quercus* are one of the most successful plant genera today. They are essentially a North temperate group with only a few species reaching South America, while none are found in Africa or Australia. There are many species

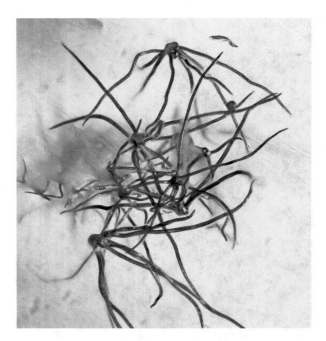

Fig. 2.6 "Oak" hairs.

Fig. 2.7 Male "oak" flower

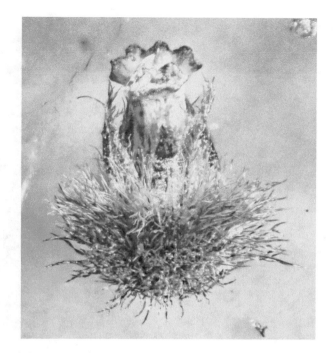

Fig. 2.8 Female "oak" flower

and varieties of *Quercus*, many of which are difficult to identify due to the ability of species of the same subgenera to hybridize. While oak trees can become huge and have one of the hardest woods in existence, this is incongruent with their extremely tiny, delicate unisexual flowers.

2.1.2   Conifers

Conifers, including cedars, cypresses, firs, douglas-firs, junipers, kauris, larches, pines, redwoods, spruces and yews, are cone-bearing seed plants found throughout the world today. While they are found in all amber forests, their remains, which include spruce needles (Fig. 2.9), cedar twigs (Fig. 2.10), and pine cones (Fig. 2.11), are most evident in Baltic amber. The seeds in one Baltic amber pine cone had started to germinate and small seedlings nearly as long as the cone had appeared (Poinar 2021) (Fig. 2.12).

Conifers depend on mycorrhizal fungi to survive. These fungi bind to the tree's roots and take in nutrients and minerals from the soil. From time to time, these mycorrhizal fungi produce fruiting bodies that resemble mushrooms. The Baltic amber mushroom, *Gerontomyces lepidotus* (Poinar 2016) (Fig. 2.13) may well have been a mycorrhizal fungus.

Fig. 2.9 Spruce needle
(*Picea* sp.)

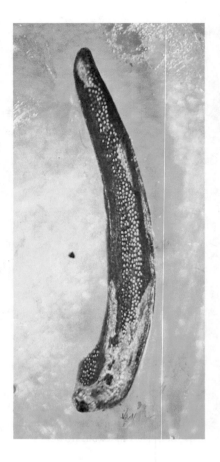

Eograminis balticus, description based on Poinar and Soreng 2021.

In moist areas of the Baltic amber forest, the coarse, green, tapering leaves of dense clumps of a tussock-forming grass (*Eograminis balticus*) were probably bent over nearly to the ground. The developing florets of this first authentic grass from Baltic amber clustered together and formed a many-flowered spikelet that was 6.8 mm long (Fig. 2.14). Due to the excellent state of preservation of the lemma, it was possible to detect stomata with well-developed subsidiary cells, parallel guard cells, silica bodies and crenulated and noncrenulated epidermal cells with long and short cells.

A fungal spore resembling those of the plant pathogenic genus *Alternaria* was found on one of the lemmas (Fig. 2.15). It may have infected *Eograminis balticus*. Holes on some of the lemmas before they had completely opened were probably made by an orthopteroid that was preserved along with the spikelet (Fig. 2.16). Similar orthopterids had been previosuy found in Baltic amber (Fig. 2.17).

Eograminis balticus resembles members of the extant genus *Molinia*, a moor grass with 2 species mostly confined to Europe, especially around the Baltic sea.

Fig. 2.10 Cedar twig
(*Thuja* sp.)

Moor grass occurs in bogs, heathlands and moorlands but also ranges from coastal to subalpine habitats, so it may have been present in the Baltic amber forest during its entire existence. Today moor grass is used in landscaping as a tufted grass and border cover. Fossils of sedges, including a species of the genus *Rhynchospora* Vahl were also recently characterized from Baltic amber (Sadowski et al. 2016).

2.1.3   Palm Flowers in Baltic Amber

***Phoenix* sp.**, description based on Poinar 2002.

Several genera of palms were previously described from Baltic amber, including *Phoenix* palms (Czeczott 1961). Based on their flowers in amber, *Phoenix* palms were fairly abundant in the Baltic amber forest. The male *Phoenix* floret that is just opening is 4.9 mm in length and has 3 sepals connnate at the base, 3 petals and 6 stamens (Fig. 2.18). It is possible that the stamens of another palm also came from a *Phoenix* floret (Fig. 2.19). Today the cultivation and sale of dates from *Phoenix* palms is a world agricultural industry and millions of tons are produced annually.

Fig. 2.11 Baltic amber pine cone

The presence of several palm lineages in Baltic amber supports the contention that the Baltic forest had a variable climate. Today, palms form a common sight in outdoor landscapes in tropical, subtropical and even warm-temperate climates around the world. Most palm flowers are unisexual with 3 sepals and 3 similar petals. Male flowers have 6 or more stamens in two whorls while female flowers have three carpels. The flowers are small and the petals mostly white. It is normally the sepals that are colored to attract pollinators. Pollinating insects may be lured by highly fragrant or even fetid odors produced by the large inflorescenses.

An assortment of insects seek out palms. Aside from the quite spectacular palm weevils and rhinoceros beetles that today ravage coconut and oil palms around the world, various caterpillars feed on the underside of palm leaves. To protect themselves from lizards that clamber over palm leaves in search of a meal, some of these

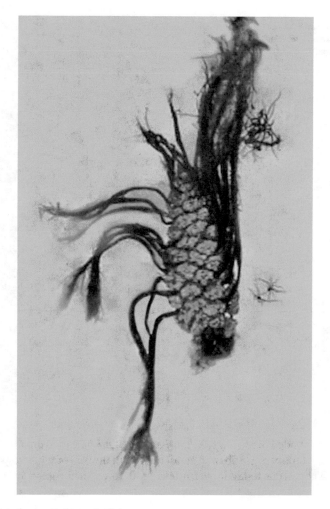

Fig. 2.12 Germinating Baltic amber pine cone

caterpillars have evolved sharp, poisonour spines over their bodies (Poinar and Vega 2019) (Fig. 2.20).

Carpantholithes berendtii, description based on Poinar et al. 2016a.

All of the cuplike flowers of *Carpantholithes berendtii* we found in Baltic amber were past flowering, had already set seed and are regarded as fruits or capsules with 3 carpel segments. They had 5 sepals, a short pedicel, a superior ovary with a short style and a globose fruiting capsule (Figs. 2.21 and 2.22). The thick walls of the hardened ovary showed different stages of opening and exposing the seeds (Fig. 2.23). The seeds were 2 mm long, ovoid and with a pitted epidermis.

Fig. 2.13 Mushroom
(*Gerontomyces lepidotus*)
in Baltic amber

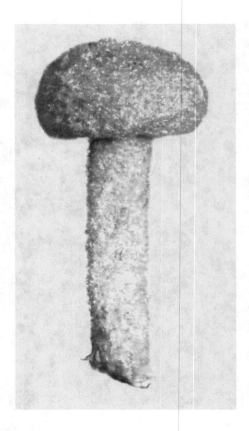

It was concluded that *Carpantholithes berendtii* is an extinct genus of the Clethraceae, a family of shrubs and small trees with flowers of white to yellow pinkish petals. They occur today in warm-temperate to tropical forests in Asia and the Americas.

Maladenodiscus acanthinus, description based on Poinar et al. 2016b.

The striking features of *Maladenodiscus acanthinus*, a bisexual flower 10 mm in length, is its large recurved sepals and prominent spine-covered ovary (Fig. 2.24). One can assume that these spines serves as a deterrent to herbivorous insects that attempted to feed on the seeds. While petals are absent, the 4 large recurved sepals unite at the base to form a large glandular nectar disc that surrounds the spiny, egg-shaped ovary. A smooth, stout style with a lobed stigma emerges from the top of the ovary (Fig. 2.25). The ring of filament scars at the base of the pistil indicate that some 40 stamens were originally present. A possible pollinator could have been a bee lineage (*Electrapis* sp.) ancestral to our honeybee (Fig. 2.26). This social bee probably nested in hollows of trees or in the ground, similar to those of our wild honeybees.

Fig. 2.14 *Eograminis balticus* spikelet

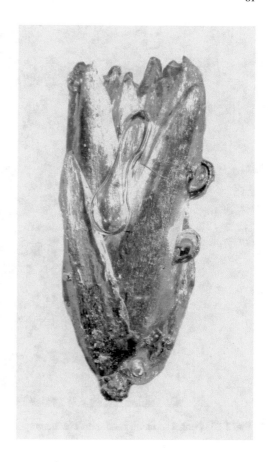

Maladenodiscus acanthinus had never been reported previously in Baltic amber and it is not possible to place it in any known family.

Succinanthera baltica, description based on Poinar and Rasmussen 2017.

While there is evidence of orchids in all of the Tertiary amber forests (Poinar 2021), a pollinarium decribed as *Succinanthera baltica* attached to a leg of a Baltic amber fungus gnat (Fig. 2.27) is the oldest fossil record of this plant family. This first orchid consists of a round pollinarium containing a sticky appendage called a visidium that attaches to insects. The fungus gnat was probably attempting to obtain nectar from the orchid flower when the pollinarium became attached to its hind leg (Fig. 2.28). Many insects distribute orchid pollinia and while fungus gnats may not be frequent visitors, they are known to pollinate present day orchids.

Based on the wide diversity of orchids found in tropical and warm-temperate forests today, the habitat of *Succinanthera baltica* could have been an epiphyte growing on rocks and branches. Assigned to the section that contains Vanda orchids,

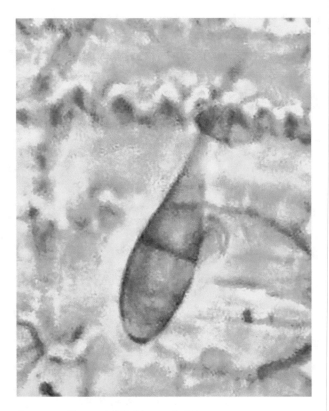

Fig. 2.15 Fungal spore on *Eograminis balticus* spikelet

the flower spikes could have had up to 30 small, spirally arranged, inconspicuous flowers hanging down from the leaves. Or they could have had large red- to orange colored blooms on pendant flower spikes up to 10 feet long arising from the leaf bases. No matter how small, the festoons of delicate colored flowers would have ornamented the murkiest of forests. With their epiphytic habit, Vanda orchids require large amounts of water so they are most common in tropical rainforests. Living as epiphytes on other plants, such as tree limbs, also brings them closer to sunlight and specific pollinators.

2.2   Summary of Newly Discovered Baltic Amber Flowers

Aside from the great diversity of angiosperm flowers that have been described in Baltic amber and placed in 38 extant families (Czeczott 1961; Caspary and Klebs 1907; Caspary 1872; Conwentz 1886), three additional genera are added in the present work; a grass, *Eograminis balticus*; an orchid, *Succinanthera baltica* and a

Fig. 2.16   Immature orthopteroid on *Eograminis balticus*

Fig. 2.17   Baltic amber orthopteroid

Fig. 2.18 Baltic amber *Phoenix* palm flower

spiked ovary flower, *Maladenodiscus acanthinus*. The latter species is the only Baltic amber flower that could not be assigned to an extant family. Presently about half of the described flowers in Baltic amber have been assigned to extant genera, but as extinct species (Czeczott 1961) (Table 2.1).

Based on the various types of vegetation preserved in Baltic amber, a possible scenario of the original amber forest some 35–40 mya would have been a distinct canopy layer composed of Kauri trees (*Agathis* spp.), pines (*Pinus* spp.), spruce (*Picea* spp.), cedars (*Thuja* spp.), oaks (*Quercus* spp.) and other members of the Fagaceae. The subcanopy and understory would have been represented by date palms (*Phoenix* sp.) and an assortment of other species. The shrub layer would have contained various palms as well as orchids (*Succinanthera baltica*). Moor grass (*Eograminis balticus*) and various ferns, mosses and fungi colonized the forest floor. Liverworts and lichens covered tree branches and trunks. While dinosaurs were absent, fossils of lizards, bird feathers and mammalian hairs reveal some of the vertebrates that were present at that period.

Fig. 2.19 Stamens of a
Baltic palm flower

Key to Newly Discovered Baltic Amber Flowers

1. Flowers borne in spikelets – 2

    1A Flowers not borne in spikelets – 3

2. Presence of short-long cell alternations and silica bodies in epidermal cells –
   Eograminis balticus

    2A Absence of short-long cell alternations and silica bodies in epidermal cells –
    Rhynchospora sp.

3. Flowers tiny, unisexual, wind-borne, lacking petals – 4

    3A Flowers not as above – 5

4. Flowers with 5–6 sepals partially fused at base and usually 6 stamens with long
   filaments and short, oval anthers – Fagaceae (male)

    4A Flowers with sepals partly covered with long trichomes and 3 carpels with 3
    short, wide styles – Fagaceae (female)

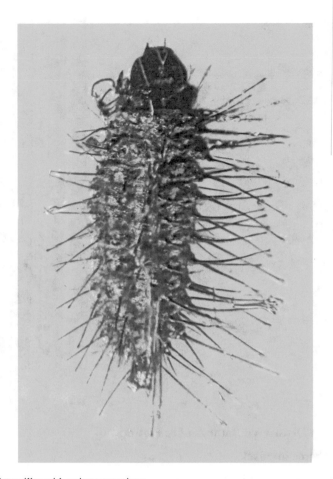

Fig. 2.20 Caterpillar with poisonous spines

5. Flowers with 3 sepals, 3 petals and 6 stamens in two whorls – *Phoenix* sp.

   5A Flowers not as above – 6

6. Flowers forming pollinaria with pollen-filled pollinia – *Succinanthera baltica*

   6A Flowers not as above – 7

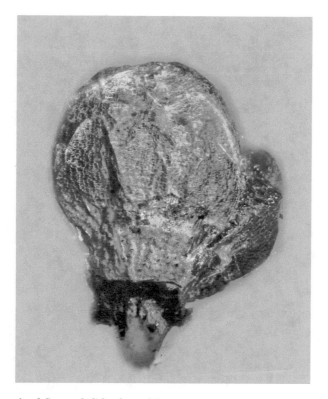

Fig. 2.21 Capsule of *Carpantholithes berendtii*

7. Mature flower with 4 large curved sepals and egg-shaped ovary covered with spines – *Maladenodiscus acanthinus*

   7A Mature flower with 5 basal sepals, superior ovary with short style and 3-valved capsule bearing 3 rounded seeds – *Carpantholithes berendtii*

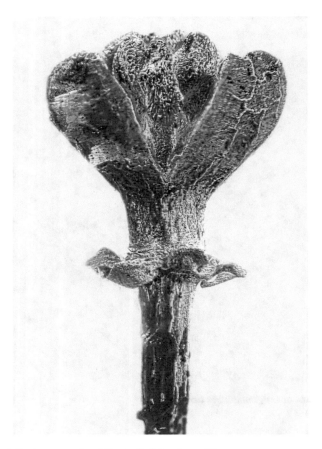

Fig. 2.22 Partially open capsule of Carpantholithes berendtii

Fig. 2.23 Completely opened capsule of *Carpantholithes berendtii* with 3 carpel segments exposing 3 seeds

Fig. 2.24 Flower of *Maladenodiscus acanthinus*

Fig. 2.25   Tapered style of *Maladenodiscus acanthinus*

Fig. 2.26   *Electrapis* sp. in Baltic amber

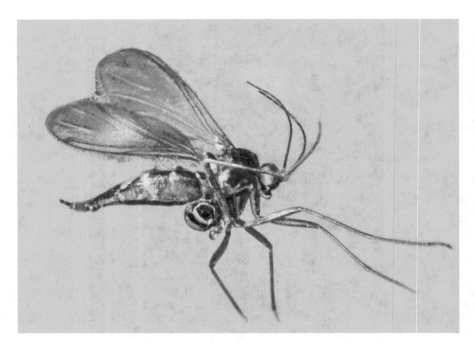

Fig. 2.27   Fungus gnat with pollinarium of *Succinanthera baltica*

Fig. 2.28   Pollinarium of
Succinanthera baltica

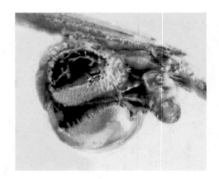

Table 2.1 Additions to previous records of Baltic amber angiosperms (N = 6)

Flower	Systematic placement	References
Carpantholithes berendtii	Clethraceae	Poinar et al. (2016a)
Eograminis balticus	Poaceae	Poinar and Soreng (2021)
Maladenodiscus acanthinus	Unknown	Poinar et al. (2016b)
Oaks	Fagaceae	Sadowski et al. (2020) Present work
Phoenix sp.	Arecaceae	Poinar (2002)
Rhynchospora sp.	Cyperaceae	Sadowski et al. (2016)
Succinanthera baltica	Orchidaceae	Poinar and Rasmussen (2017)

References

Caspary R (1872) Pflanzliche Reste aus der Bernsteinbildung. Privatsitzung am 4. October. Schriften der Königlichen Physikalisch-Ökonomischen Gesellschaft zu Königsberg 13: 17–18

Caspary R, Klebs R (1907) Die Flora des Bernsteins und anderer fossiler Harze des ostpreußischen Tertiärs, Band I. Abhandlungen der Königlich Preußischen Geologischen Landesanstalt, Neue Folge 4: Berlin: 1–181

Conwentz HW (1886) Die Flora des Bernsteins, vol 2. Die Angiosperms des Bernsteins. Wilhelm Engelmann, Gdansk

Conwentz HW (1890) Monographie der baltischen Bernsteinbäume: vergleichende Untersuchungen über die Vegetationsorgane und Blüten, sowie über das Harz und die Krankheiten der baltischen Bernsteinbäume. Engelmann, Danzig, pp 1–151

Czeczott H (1961) The flora of the Baltic amber and its age. Prace Muzeum Ziemi (Paleobotaniczne) Warszawa 4:119–145

Larsson SG (1978) Baltic Amber- a palaeobiological study. Scandinavian Science Press, Klampenborg

Legalov AA (2016) Two new genera and four new species of fossil weevils (Coleoptera: Curculionoidea) in Baltic amber. Entomologica Fennica 27:57–69

Poinar GO Jr (2002) Fossil palm flowers in Dominican and Baltic amber. Bot J Linnean Soc 139:361–367

Poinar GO Jr (2016) A gilled mushroom, *Gerontomyces lepidotus* gen. et sp. nov. (Basidiomycota: Agaricales), in Baltic amber. Fungal Biol 120:1–4

Poinar GO Jr (2021) Precocious germination of a pine cone in Eocene Baltic amber. Hist Biol. https://doi.org/10.1080/08912963.2021.2001808

Poinar GO Jr, Hess RT (1982) Ultrastructure of 40-million-year-old insect tissue. Science 215:1241–1242

Poinar GO Jr, Rasmussen FN (2017) Orchids from the past, with a new species in Baltic amber. Bot J Linnean Soc 183:327–333

Poinar GO Jr, Soreng RJ (2021) A new genus and species of grass, *Eograminis balticus* (Poaceae: Arundinoideae), in Baltic amber. Int J Plant Sci 182. https://doi.org/10.1086/716781

Poinar GO Jr, Vega FE (2019) Poisonous setae on a Baltic amber caterpillar. Arthropod Struct Dev 51:37–40

Poinar GO Jr, Chambers KL, Brown AE (2016a) *Carpantholithes*, a restored generic name for Eocene fossils in Baltic amber representing an extinct lineage of Clethraceae. J Bot Res Inst Texas 10:147–153

Poinar GO Jr, Chambers KL, Brown AE (2016b) *Maladenodiscus acanthinus*, a new genus and species of fossil flower in Baltic amber. J Bot Res Inst Texas 10:443–447

Sadowski E-M, Schmidt AR, Rudall PJ, Simpson DA, Gröhn C, Wunderlich J, Seyfullah LJ (2016) Graminids from Eocene Baltic amber. Rev Palaeobot Palynol 233:161–168

Sadowski E-M, Schmidt AR, Denk T (2020) Staminate inflorescences with *in situ* pollen from Eocene Baltic amber reveal high diversity in Fagaceae (oak family). Willdenowia 50:405–517

Sokoloff DD, Ignatov MS, Remizowa MV, Nuraliev MS, Blagoderov V, Garbout A, Perkovsky EE (2018) Staminate flower of *Prunus* s.l. (Rosaceae) from Eocene Rovno amber (Ukraine). J Plant Res 131:925–943

Weitschat W, Wichard W (2002) Atlas of plants and animals in Baltic amber. Verlag Dr. Friedrich Pfeil, Munich

Wolfe AP, McKellar RC, Tappert R, Sodhi RNS, Muehlenbachs K (2016) Bitterfeld amber is not Baltic amber: three geochemical tests and further constraints on the botanical affinities of succinite. Rev Palaeobot Palynol 225:21–32

Chapter 3
Dominican amber Flowers

Abstract Based on direct and indirect evidence, 38 angiosperm flowers in over 20 plant families enhanced the tropical moist Dominican amber forest with their beauty, grace and mysterious qualities. While 20 flowers in Dominican amber can be placed in current families and genera, 5 belong to unknown families, showing that extinctions have occurred at the species, genus and family levels over the past 20–30 million years. None of the Dominican amber flowers represent present day species. Here we have direct evidence of pollinators, such as stingless bees entrapped in the stamens of mimosoids, as well as herbivores that have nibbled on petals. Stingless bees carrying pollinia also show some of the orchid diversity that was in the forest at that time.

The landmass that contained the Dominican Republic amber forest was formed some 100 mya by volcanic activity that occurred along a marine shelf in the Caribbean between North and South America. This volcanic eruption resulted in the formation of the "Proto-Greater Antilles". This landmass rose above the sea some 60 mya and slowly began drifting eastward. Life forms from both North and South America began to arrive soon after its emergence and continued to appear during the movement and break up of the Proto-Greater Antilles landmass into the various islands now known as Jamaica, Cuba, Puerto Rico and Hispaniola.

Plants and animals in the Dominican amber forest first appeared in the time span between the emergence of the land mass from the sea some 60 mya and its arrival in the present location about 25 mya. The forest revealed its unmistakable presence through the formation of what has been called "new world amber", the majority of which occurs in the northern mountain ranges of the Dominican Republic on the island of Hispaniola.

It is probable that the tropical moist Dominican amber forest on the island of Hispaniola thrived in all its glory in essentially the same geographical location and under similar climatic conditions that exist today. This is why we see many flowers similar to those found in the Caribbean today. However, while the great majority of the families of the Dominican amber flowers may be still extant, all the flowers found thus far in Dominican amber are now extinct at the species, and often, the generic level. What Dominican amber flowers show us is that angiosperms in the Dominican amber forest were more extensive than they are at present.

The Dominican amber forest has been categorized as a tropical moist forest unlike any other found in the world today (Poinar and Poinar 1999). During the Pliocene-Pleistocene global cooling period, many of the tropical biota of the forest, such as stingless and orchid bees, *Mastotermes* termites, as well as various plants, including *Hymenaea protera* that produced the amber, became extinct. Refugia were inadequate for many life forms and today, only related lineages occur in other parts of the Neotropical realm.

Just when the resin-bearing legume tree (*Hymenaea protera*) that produced the amber first appeared is unknown, so we don't know how much of the original forest is represented in Dominican amber. When we recently searched for amber in the Dominican Republic, we encountered some of the same invertebrate pests that were present when the amber forest was thriving. These include mosquitoes carrying malaria, kissing bugs vectoring trypanosomatids and ticks supporting a range of pathogens.

The majority of Dominican amber comes from mines in the Cordillera Septentrional of the Dominican Republic (Fig. 3.1). Perhaps the most famous mine is La Toca (Fig. 3.2), followed by the Palo Alto mine (Fig. 3.3) since they contain very fossiliferous clear amber. The dating of Dominican amber is still controversial with the latest proposed age of 20–15 mya based on foraminifera (Iturralde-Vinent and MacPhee 1996, 2019) and the earliest as 45–30 mya based on coccoliths (Cêpek in Schlee 1990). In addition, Dominican amber is secondarily deposited in sedimentary rocks, which makes a definite age determination difficult (Poinar and Mastalerz 2000). A range of ages for Dominican amber is possible since the amber is associated with turbiditic sandstones of the Upper Eocene to Lower Miocene Mamey Group (Draper et al. 1994).

Hymenaea protera, description based on Poinar 1991

Dominican amber was produced from resin of the leguminous tree, *Hymenaea protera*, which is now extinct. Based on the growth habits of extant species of *Hymenaea* in other parts of the world, *Hymenaea protera* would have extended up into the canopy layer. The flower clusters near the top of the tree would have had clouds of insects swirling around them in search of pollen and nectar. The flowers would have matured quickly and soon after they opened, there would have been a continuous series of white showy petals wafting downward. Some ended their descent in spider webs while others fell together with herbivores and pollinators.

The flowers of *Hymenaea protera* are large, reaching 20 mm in length, with 4 spirally arranged elliptical sepals (Fig. 3.4), 3 stalked, heart-shaped, showy, early caducous petals (10–15 mm long) and 2 min, persistent petals (under 1.5 mm in length). The function of the small persistant petals, which are glabrous and elliptical in outline (Fig. 3.5), is unknown. The showy petals are tan-colored, glabrous, strongly clawed with a central midrib and cordate to reniform base (Fig. 3.6).

The ovary is rhomboidal to oblong and completely glabrous except for elongate hairs at the base and along one lateral margin. As the ovaries mature, their surface

Fig. 3.1 Area of amber mines in the Cordillera Septentrional of the Dominican Republic

Fig. 3.2 Worker preparing to enter La Toca amber mine in the Cordillera Septentrional

Fig. 3.3 Collecting amber at the Palo Alto mine in the Cordillera Septentrional

becomes dark brown and verrucose and the style, bearing a small capitate stigma, becomes outstretched (Figs. 3.4 and 3.5). The fruit is an elongate pod but it is difficult to find mature pods in amber. The stamens have long, thin filaments with rod-shaped anthers. The triporate pollen grains, masses of which have been released in some stamens that ended up in amber (Fig. 3.7), are spherical to sub-prolate and range from 27 to 34 µm in diameter. Their shape and size make them easy to recognize on the bodies of stingless bees (Fig. 3.8).

The leaves of *Hymenaea protera* are paired with both leaflets identical and attached opposite each other. However, the leaf blades are unequal at the base and rounded more on the outside than the inside (Fig. 3.9). Earlier studies showed evidence of extracting, amplifying and sequencing chloroplast genes from a leaflet of the Dominican amber tree (Poinar et al. 1993).

Based on the morphology of *Hymenaea protera,* the closest known extant species is *Hymenaea verrucosa* from Malagasy and Mauritius. It is likely that the genus *Hymenaea* originated when the South American and African plates formed a

Fig. 3.4 *Hymenaea protera* flower with sepals and developing ovary

common land mass. As the continental plates separated, the distribution of *Hymenaea* became disjunct with populations remaining on both continents (Poinar 1991).

Trochanthera lepidota, description based on Poinar et al. 2008b

While many Dominican amber flowers can be assigned to present day families and genera, some could not. One of these mystery flowers is *Trochanthera lepidota*. In the amber forest, the flowers of this leafless plant probably would have emerged from a subterranean tuber among decaying vegetation. *Trochanthera lepidota* is a male plant covered with numerous small flowers (Fig. 3.10). Each flower, which is slightly over 7.0 mm in width, has 3–4 appressed stamens with short, stout filaments and anthers (Fig. 3.11). The protruding numerous stamens gives the flower its rough appearance. All of the anthers were open and shedding pollen.

It is likely that *Trochanthera lepidota* was a parasite on the roots of adjacent shrubs and trees, similar to the habits of ground cones of the family Orobanchaceae or spike-of-dragons of the family Balanophoraceae.

Fig. 3.5 Minute persistent petal at base of ovary with adjacent stingless bee

Brevitrimaris arcuatus, description based on Chambers and Poinar 2016b

Another flower in the Dominican amber forest that could not be placed in a current family is the bisexual *Brevitrimaris arcuatus*. This small 3.3 mm wide flower has its floral parts in threes; 3 sepals, 3 petals and 3 stamens, which are characters of monocots, a group that includes orchids, palms and grasses (Fig. 3.12). The curved anthers in the center of the flower mask surround the style and stigma (Fig. 3.13). The sepals are covered with long hairs while the outer surface of the petals is covered with rounded cells, giving them the appearance of cobblestones. The presence of round pollen grains on the anthers and stigma show that the flower was in full bloom when it entered the resin. While small stingless bees could have been pollinators, the proximity of the anthers to the stigma suggests that self pollination might also have occurred.

Brevitrimaris arcuatus has features of members of the monocot family Haemodoraceae. This family consists mostly of perennial herbs with parallel-veined

Fig. 3.6 Large, showy
petal of *Hymenaea protera*

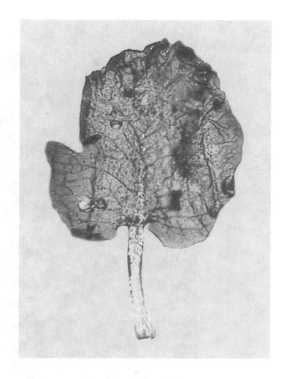

long leaves and slender stems bearing small clusters of bisexual 6-tepaled flowers
with 1–6 stamens.

Phyrtandrus pentalepidus, description based on Chambers and Poinar 2016c

This small male *Phyrtandrus pentalepidus* flower, with a width of only 2.7 mm,
is one of the strangest flowers in Dominican amber (Fig. 3.14). The flower possesses
a combination of normal, unusual and abnormal characters that make it impossible
to place in any current plant family. The outer ring of 5 stamens just above the 5
broad sepals are functional and the anthers have produced large round, monosul-
cate, pollen grains from 24 to 29 microns in diameter (Fig. 3.15).

However in the second or middle ring, there are no normal stamens, just two
infertile, spherical, malformed structures that can be considered staminodes. These
infertile staminodes are spherical in shape and composed of a tangled mass of trichomes (Fig. 3.16). Whether these trichomes, which produce no pollen grains, are
modified filaments or abnormal germinating intertwined pollen grains is unknown.
In the innermost stamen ring are two normal stamens and one elongate sterile stamen, showing how fertile and infertile stamens can be produced side by side. While
the female structures (ovary, style, etc.) are absent, perhaps enough female hor-
mones are present to cause some of the male stamens to be infertile and malformed.
In looking for some possible function of the spherical staminodes with their tangled
hairs, perhaps they were secreting nectar or releasing some odor to attract

Fig. 3.7 Anther of *Hymenaea protera* shedding pollen

Fig. 3.8 Stingless bee bearing large, round pollen

Fig. 3.9 Leaflet of
Hymenaea protera

pollinators. A few aberrant pollen grains from these abnormal stamens possessed numerous minute cones over their surfaces.

Phyrtandrus pentalepidus has a mixture of lily (monocot) and dogwood (eudicots) features, possible arising from a lineage that dates back to the union of these two major groups of flowering plants. The flower could not be assigned to any known family and it is difficult to determine what insects served as pollinators. However, some stingless bees in Dominican amber are covered with large, round pollen grains matching the size and shape of those from *Phyrtandrus pentalepidus*.

Distigouania irregularis, description based on Chambers and Poinar 2014a

Another male flower in Dominican amber that could not be placed in a current genus is the small 3.4 wide male *Distigouania irregularis*. A curious feature of this flower is that while the sepals are irregular in shape, they are larger than the petals (Fig. 3.17). Glandular appendages that probably secreted nectar are positioned at the base of the sepals. The flower has an irregular corolla with 4 rhombic-lanceolate, spreading, sepal-like, petals and a 5th shorter, erect, and slightly cupped petal. Its

Fig. 3.10 *Trochanthera lepidota* flower

Fig. 3.11 Stamens of
Trochanthera lepidota

Fig. 3.12   Flower of *Brevitrimaris arcuatus*

Fig. 3.13   Curved anthers surrounding style of *Brevitrimaris arcuatus*

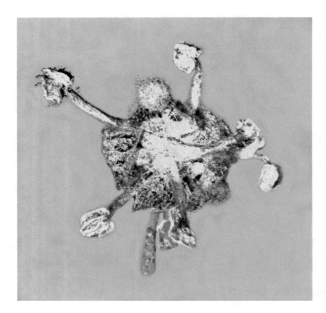

Fig. 3.14 *Phyrtandrus pentalepidus* flower

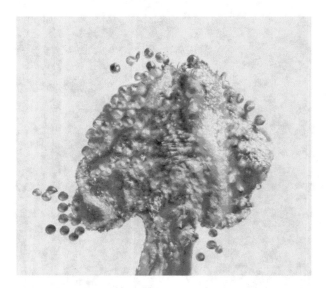

Fig. 3.15 Anther and pollen of *Phyrtandrus pentalepidus*

Fig. 3.16 Dysfunctional anther of *Phyrtandrus pentalepidus*

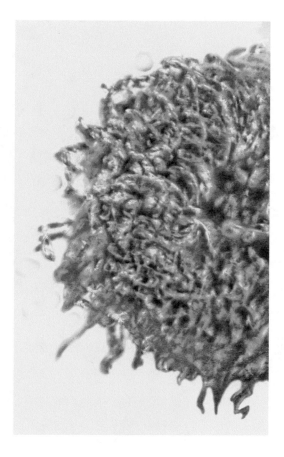

specific name is based on the irregular shape of the 5 petals. There are five dimorphic stamens with 4 producing pollen on erect filaments opposite the spreading petals and one that adjoins the cupped petal with an unopened larger anther. Another unusual feature is that in the center of the flower, where the ovary would be located if the flower was bisexual, is a hypanthial disc with a cluster of tangled hairs (Fig. 3.18).

Distigouania irregularis was assigned to the Rhamnaceae family but could not be placed in any current genus, although it has some features of *Gouania*. Members of this genus are shrubs and lianas found in tropical and subtropical forests in Mexico, Central America and the Caribbean. The Caribbean species, which is a liana, is used to make a commercial toothpaste and the local inhabitants chew stems and twigs of the plant to clean their teeth.

Fig. 3.17 *Distigouania irregularis* flower

Fig. 3.18 Tangled hairs in center of *Distigouania irregularis* flower

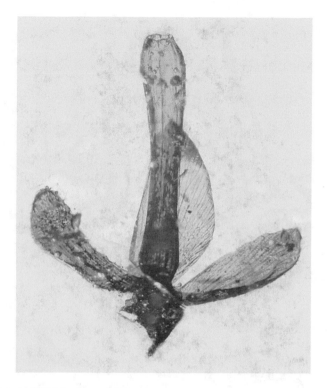

Fig. 3.19   Flower of *Dasylarynx anomalus*

Dasylarynx anomalus, description based on Poinar and Chambers 2015a

Dasylarynx anomalus is represented by two separate flowers in different amber pieces. The holotype specimen is complete while the paratype is only a partial specimen that had a portion of the fused petal tube polished away, revealing the inner features of this strange bisexual flower. The most striking feature of the flower is the extended "floral tube" that is formed by the fusion of the petals (Fig. 3.19). The sexual organs, involving some 9 stamens and a stalked ovary with a long single style are located at the bottom of the "floral tube". This is one of the largest flowers found in Dominican amber, being nearly 16 mm in total length.

Especially interesting is the coating of long slender, pointed hairs (trichomes) on the inner walls of the floral tube of *Dasylarynx anomalus* (Fig. 3.20). It is possible that these upward pointing hairs served to block ants and other insects from reaching nectar deposits. The only way for pollinators to reach the spherical pollen grains and stigma at the base of the tube would be to enter through the opening at the top of the tube and travel the hazardous route to the bottom against the upward pointing

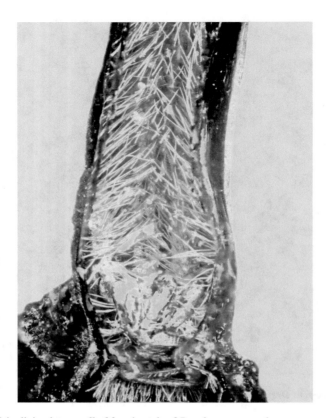

Fig. 3.20 Hairs lining inner wall of fused petals of *Dasylarynx anomalus*

trichomes. Or they could remain on the top and insert their probosces or beaks, if hummingbirds were attracted, down through the hairs to reach the nectar and pollen at the bottom. Other possible pollinators are lepidopterans, such as sphinx moths or riodinid butterflies that could extent their probosces down the tube to obtain nectar at the bottom as long as the thick coating of hairs would not hinder them. The 3 equal sepals, 3 fused petals and 3 whorls of 3 stamens align *Dasylarynx anomalus* with monocots but the flowers do not resemble those of any modern genus or family.

Pseudhaplocricus hexandrus, description based on Poinar and Chambers 2015c

As previously mentioned, while most of the flowers in Dominican amber can be placed in current families, many cannot be placed in extant genera. This is the case with the small, 3.3 mm wide, male flower of *Pseudhaplocricus hexandrus*. While dominated by the 6 stamens, the flower has 3 broad sepals that extend behind the staminal ring, however its 3 petals have fallen (Fig. 3.21). With the petals, the flower would probably be closer to 9 mm in width. While there is no ovary or pistillode present, what is so noticeable are the 6 outstretched stamens with their roundish anthers that were shedding pollen when the flower fell into the resin (Fig. 3.22).

Fig. 3.21   Male flower of *Pseudhaplocricus hexandrus*

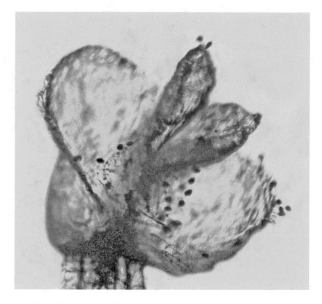

Fig. 3.22   Anther with pollen of *Pseudhaplocricus hexandrus*

Features of this fossil flower place it in the Day flower family (Commelinaceae). It would probably have been an herb with fairly long, slender stems bearing parallel-veined leaves and small clusters of whitish flowers. It would have made a pleasant, but unassuming bloom in flower gardens.

If *Pseudhaplocricus hexandrus* was similar to present day Commelinaceae, it may not have produced nectar to attract pollinators since the only reward Commelinaceae flowers offer is pollen. However the center of the fossil flower is depressed, suggesting that it could have contained some nectar produced by subsurface glands. Insects, like bees and muscoid flys (Fig. 3.23) that pollinate Commelinaceae flowers today, could have used the sepals as landing platforms when they fed on the pollen.

Fig. 3.23 Muscoid fly in Dominican amber

3.1 Mimosoideae (Fabacese)

The striking flowers of mimosoid herbs, shrubs, trees and vines must have filled the Dominican amber forest with a variety of colors. Their numerous stamens radiated out like wire strands, sometimes intertwined with one another (Fig. 3.24). Quite a few species left flowers, leaves and even a few seeds behind in amber.

Campopetala dominicana, description based on Poinar and Chambers 2016

The two bisexual mimosoid flowers of the new genus and species, *Campopetala dominicana*, range from 4.8 to 5.1 mm in length. The flowers are characterized by a cup-shaped calyx, 5 reflexed petals, 20–30 stamens with extremely long filaments (8–11 mm), a short style and transverse anthers with protruding pollen sacs at either end (Figs. 3.25 and 3.26). Apical anther glands were lacking. The long filaments form a tangled mass that extend above the petals. The structure and preferred habitat of the mature plant are unknown.

Entada hispaniolae, description based on Poinar and Chambers 2016

Several bisexual flowers of a new species of the present day genus *Entada* were recovered from Dominican amber. These sessile flowers with a small 5-part calyx

Fig. 3.24 Mimosoid flower with extended stamens

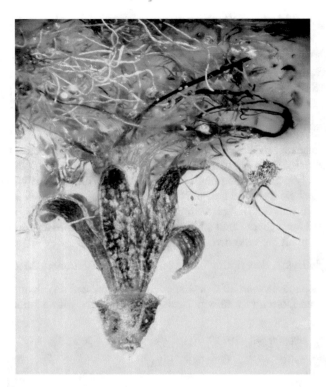

Fig. 3.25 *Campopetala dominicana* flower

Fig. 3.26 Pollen sacs at opposite ends of *Campopetala dominicana* anther

Fig. 3.27 *Entada hispaniolae* flower in Dominican amber

bearing sepals with minute teeth are only 5 mm long. The 5 long, slender petals form a closed basal tube around the 10 stamens (Fig. 3.27). The anthers are quite distinct with each one possessing a small, spherical gland attached only to one end (Fig. 3.28). When the flowers first open, the glands probably secrete attractants for pollinators, since they detach shortly afterwards. The slender style is capped with a blunt-tipped stigma.

Entada is basically an Old World genus of woody lianas that can become quite tall. A few are armed with prickles while others have smooth limbs and foliage. Only four out of the 30–40 current species occur in the Americas, where they thrive in lowland rain and streamside forests and coastal thickets, often in sandy soils. The Neotropical species are known for their extremely large seed pods, which can reach over a meter in length. These large pods have been used as clubs by law enforcement agents. The seeds have been used as a substitute for coffee and also as a shampoo and laundry soap due to the their high saponin content.

Senegalia eocaribbeansis, description based on Poinar and Chambers 2016

Senegalia eocaribbeansis must have been fairly abundant in the Dominican amber forest or grew near resin-producing *Hymenaea* trees. Some amber pieces contain up to 4 small *Senegalia eocaribbeansis* flowers, each with 5 petals under

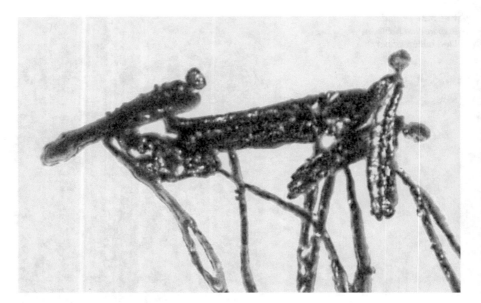

Fig. 3.28 Tiny spherical glands on only one end of the anthers of *Entada hispaniolae*

2 mm in length. Since both male and bisexual flowers occur in the same piece of amber, some plants were dioecious with both sexes on the same tree.

There can be as many as 80 stamens in a single *Senegalia eocaribbeansis* flower with the anthers barely visible to the naked eye (Fig. 3.29). Many anthers have stalked apical glands that probably produced compounds attractive to pollinators (Fig. 3.30). In bisexual flowers the ovary is flat with a round but hairless surface and the style is nearly straight with the lower part curved. Based on closely related present day *Senegalia* species, the flowers of *Senegalia eocaribbeansis* were probably yellow, the leaves pinnately compound and the pods were somewhat long and contained up to 5 flattened roundish seeds.

Several stingless bees are still entangled in the stamens so we know they were pollinators that crawled though the mass of filaments in search of pollen and nectar (Fig. 3.31). Aside from a stingless bee preserved along with the blooms, a small hemipteran that had its mouthparts inserted in an anther was also entrapped (Fig. 3.32). Gum Arabic from *Senegalia* trees in Africa is used in glue, the food industry, watercolor paints and burnt for incense.

The diverse assortment of mimosoids in the Dominican amber forest can also be determined by the various leaves and leaflets that occur in amber. The leaves of these plants are usually pinnate, with each leaf composed of a number of leaflets (pinnules) arranged on one or both sides of the stalk (petiole). The shape, size and texture of these leaflets vary considerable and are unique to different genera. The leaflets can be erect or folded inward with pointed tips (Fig. 3.33), broadly oblong with rounded tips (Fig. 3.34), elongate and slightly curved (Fig. 3.35) and with pubescent surfaces (Fig. 3.36). There are ample examples of herbivory on the

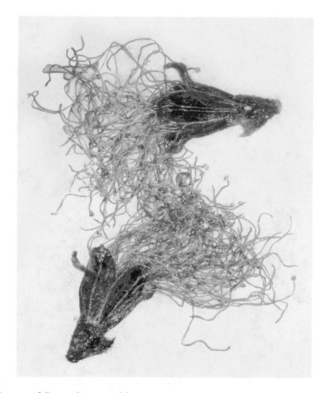

Fig. 3.29   Flowers of *Senegalia eocaribbeansis*

margin (Fig. 3.37) or surface (Fig. 3.38) of mimosoid leaflet in Dominican amber. Various herbivores feeding on the leaves attract ants, and one *Azteca* ant was apparently obtaining honeydew from aphids on one leaf (Fig. 3.39). Mimosoid seeds are scarce in amber but occasionally are found, often with insect damage (Fig. 3.40).

Ekrixanthera hispaniolae, description based on Poinar et al. 2016

In life, *Ekrixanthera hispaniolae* would probably have been a large shrubby nettle with clusters of star-shaped flowers arising from the leaf axils. Male flowers of *Ekrixanthera hispaniolae* were found in 3 separate pieces of Dominican amber. While petals are absent, flowers of *Ekrixanthera hispaniolae*, which range between 4 and 5 mm in diameter, possess 5 stamens, each subtended by a narrow sepal (Fig. 3.41). The filaments are broad and bear large bi-lobed anthers. The flower stalks are quite short so the inflorescent must have been spike-like. A hairy pistillode is in the center of the flower.

One piece with 2 separate male flowers also contained a developing fruit (achene) (Fig. 3.42) and a caterpillar (Fig. 3.43). The developing fruit is roundish, smooth-walled, lacks a stigma and is surrounded by 5 unequal sepals. Considering these features, *Ekrixanthera hispaniolae* was placed in the nettles family (Urticaceae).

Fig. 3.30   Stalked apical anther gland of *Senegalia eocaribbeansis*

Several insects were in the amber pieces with *Ekrixanthera hispaniolae*. Adjacent to one of the flowers was a worker Dolichoderinae ant (Hymenoptera: Formicidae) (Fig. 3.44) that was probably searching for nectar. Adjacent to the developing fruit were several adult gall midges (Diptera: Cecidomyiidae). Gall midge larvae are known to develop on nettles today.

Ekrixanthera hispaniolae flowers resemble those of the present day genus *Boehmeria*, which are mostly small woody trees, shrubs and herbaceous perennials. If the growth pattern of *Ekrixanthera hispaniolae* was similar to that of the Bog Hemp (*Boehmeria cyclindrica*), it would have been a meter or so in height at maturity and bear clusters of greenish-white flowers in the axils of the upper leaves.

Discoflorus neotropicus, description based on Poinar 2017c

It is rare to find a fossil flower that is preserved together with its pollinator. However, here is a milkweed flower adjacent to its termite pollinator. The bisexual

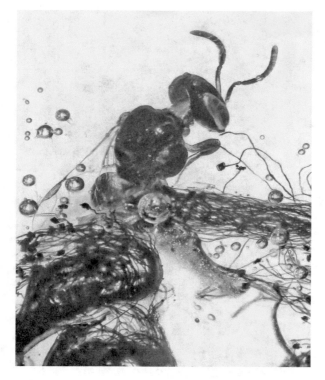

Fig. 3.31 Stingless bee pollinator of *Senegalia eocaribbeansis*

Discoflorus neotropicus is just under 3 mm in length and possesses 5 small sepals and 5 long petals (Fig. 3.45). The 5 stamens alternate with elongate scales to form a ring around the 10-lobed flattened stigmatic disk (Fig. 3.46). The anthers are fused with the stigmatic region of the gynoecium, thus forming a gynostegium.

The milkweed flower resembles those of the present day genus *Metastelma*, which are vine milkweeds with stems that twist around other plants and form dense groups. The white flowers are borne in the axils of thick glossy leaves with curved tips. Features of *Discoflorus neotropicus* established it as a new genus in the family Apocynaceae.

Today, representatives of two major plant families, the milkweeds (Apocynaceae) and the orchids (Orchidaceae), distribute their pollen in little sacs called pollinia. The pollinia with their attachment devices are known as pollinaria. In the amber adjacent to the milkweed flower is a termite (Fig. 3.47) with two large spherical pollinaria attached to its clypeus (Fig. 3.48). Termites are usually not regarded as pollinators, however members of the termite family Termitidae, to which the fossil termite belongs, are known as harvester termites. These workers collect flowers and other plant material for their nests.

Fig. 3.32 Hemipteran herbivore of *Senegalia eocaribbeansis*

It is common for present day milkweeds to secrete nectar from their stigmatic disks and it is possible that this is why the amber termite was attracted to the milkweed flower. It is likely that other insects, like butterflies and bees, were also pollinators of *Discoflorus neotropicus*. However night foraging termites may be more important pollinators than we realize.

3.2   Orchids

The presence of orchids in Dominican amber includes direct evidence from floral remains, pollinaria attached to insects, and an isolated seed. Indirect evidence is from the presence of orchid bees in various amber pieces. As mentioned previously when discussing the milkweed, *Discoflorus neotropicus*, orchids are another plant family whose flowers produce pollen in small sacs called pollinia. These pollinia are in pollinaria that become attached to the bodies of a wide range of insects by sticky

Fig. 3.33 Leaflets folded
inward with pointed tips

pads (viscidia). Finding these pollinators in amber is a rare event, but those found in Dominican amber show a sample of the variety of orchids that existed at that time.

Cylindrocites browni, description based on Poinar 2016c

The piece of amber containing the *Cylindrocites browni* pollinarium also has portions of the orchid petals, one of which is 16 mm in length, that probably produced the pollinarium (Fig. 3.49). The surface of the petals are covered with thick hairs (trichomes) (Fig. 3.50). The pollinarium, which is attached to the head of the weevil, is 2 mm long, which is longer than half the length of the insect (Fig. 3.51). The cylindrical pollinarium is ornamented with dark bars and spots (Fig. 3.52). It is attached to the weevil by a sticky roundish deposit (viscidium) at one end. The weevil probably just left the orchid flower when it fell into the resin. The pollinarium resembles those of Neotropical orchids of the subfamily Spiranthinae. These are mostly terrestrial or epiphytic forms with short leafy stems and tender leaves.

Fig. 3.34 Leaflets with
rounded tips

Globosites apicola, description based on Poinar 2016b

Bees, with their constant visits to the flowers, are the most common carriers of
orchid pollinaria. The first record of this association in Dominican amber was
Meliorchis caribea attached to the back of a stingless bee (Ramírez et al. 2007).
Meliorchis caribea was placed in the subtribe Goodyerianae of the Tribe Cranichidae
in the subfamioly Orchidoidea. In the Dominican amber forest, small stingless trigonid bees visited a wide range of flowers for nectar and pollen so it is not surpris-
ing that some would have acquired pollinaria from various orchids. The bulbous
pollinarium of *Globosites apicola*, about 0.5 mm in diameter, is quite noticeable
attached to the face of a worker stingless bee (Figs. 3.53 and 3.54). How it was
removed when the bee visited another flower of *Globosites apicola* is unknown.
Extant Neotropical orchids having a similar type of pollinaria are members of the
Oncidiinae. These are mostly epiphytic orchids pollinated by bees and wasps.

Fig. 3.35 Leaflets
elongated and slightly
curved

Rudiculites dominicana, description based on Poinar 2016b

Pollinia come in all shapes and sizes. *Rudiculites dominicana* is shaped like an old long-handled wooden spoon (Fig. 3.55). The entire pollinarium, which is slightly over 1 mm in length, has an enlarged "bowl" portion, filled with developing pollen grains that projects above the bee's head (Fig. 3.56). The sticky adhesive visidium extends up on the handle, attaching it securely to the bee's head.

   Rudiculites dominicana was placed in the Neotropical subfamily Spiranthinae, which includes terrestrial and epiphytic orchids with small to medium sized flowers. Pollination of this group is normally by bees that are attracted to nectar deposits. Today, native trigonid bees are extinct in the Greater Antilles, which may explain why *Rudiculites dominicana* and many other flowers in Dominican amber are absent in the present flora.

Fig. 3.36 Leaflets
pubescent

Mycophoris elongatus seed with *Synaptomitus orchiphilus* fungus, description
based on Poinar 2017a, b, 2021

Plants in amber can also be identified by reproductive structures such as fruits
and seeds. The 1.3 mm long seed of *Mycophoris elongatus* (Fig. 3.57) possessed the
basic features of extant orchid seeds, including an associated fungus, *Synaptomitus
orchiphilus* (Fig. 3.58). Most, if not all, orchid seeds germinate more successfully
when invaded by certain mycorrhizal fungi that provide nutrients needed for the
orchid embryo to continue its development. After entering the seed, these basidio-
mycete fungi develop intracellularly, going through various developmental stages
inside their "hosts". Some, like the fossil *Synaptomitus orchiphilus*, even form
spores inside the developing orchid seeds.

Fig. 3.37 Herbivore
damaged leaflet margin

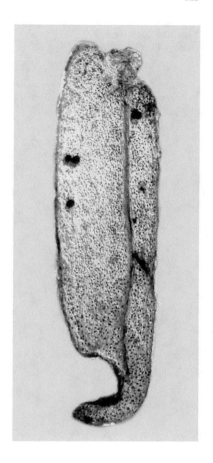

The basal sector of *Mycophoris elongatus* possesses suspensor cells and developing embryo cells undergoing division on the lower end. The upper end contains extended testa cells (Fig. 3.57). In the embryonic cells of *Mycophoris elongatus* are developing hyphae of *Synaptomitus orchiphilus*, along with pelotons. These loose aggregations of hyphae that form irregular clumps that are eventually digested by the orchid cells are characteristic of orchid seeds hosting symbiotic fungi (Poinar 2017b, 2021).

Fig. 3.38 Herbivore
damaged leaflet surface

3.3   Orchid Bees

Paleoeuglossa melissiflora, description based on Poinar 1998

Aside from direct evidence of orchids in Dominican amber, there are indirect ways
to determine their presence, like locating specific pollinators in amber. Many orchids
are dependent on male orchid bees to pollinate their flowers during their visits in
search of aromatic substances that are then converted into female sexual attractants.
Female orchid bees possess long "tongues" that are used to obtain nectar from the
base of floral tubes of various flowers. They also collect resin for nest formation.
Some female orchid bees entombed in Dominican amber (Figs. 3.59, 3.60, and
3.61) were apparently attempting to gather resin. These metallic appearing bees can
be quite large, some reaching over 16 mm in length (Poinar 1998).

Fig. 3.39 *Azteca* ant that had been attending aphids on mimosoid leaf

As with stingless bees, endemic orchid bees are now absent from Hispaniola. Their disappearance could account for the extinction of various orchids that depended on these bees for pollination.

Swietenia dominicensis, description based on Chambers and Poinar 2012b

The curious bisexual flower of *Swietenia dominicensis*, just under 4 mm in length, has a unique central pitcher-shaped floral tube composed of 10 stamens alternating with elongate floral appendages (Fig. 3.62). This combined ring of stamens and appendages surrounds the ovary, which has a very wide disc-shaped stigma covered with minute pores (Fig. 3.63). The function of these pores is unknown but they could be emitting fragrances or secreting nectar to attract pollinators.

Fig. 3.40 Herbivore damaged mimosoid seed

Although the sepals are very small, the petals are fairly long and in life may have been greenish-white, similar to the color of its modern descendant, the West Indian Mahogany tree (Meliaceae). This latter tree, which is the national tree of the Dominican Republic, is also found today in Florida, the Bahamas, Cuba and Jamaica. It originally supplied most of the world's mahogany that was used in ship building, furniture making, and musical instruments. Now, due to the removal of so many West Indian mahogany trees, most commercial mahogany is supplied by related tree species overseas.

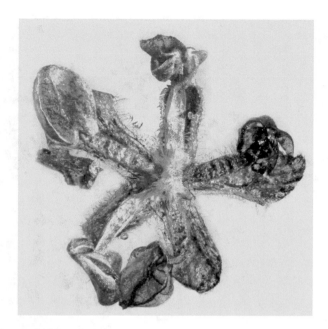

Fig. 3.41   Flower of *Ekrixanthera hispaniolae*

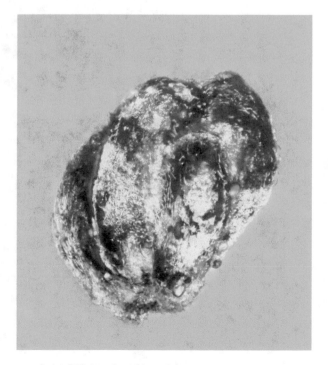

Fig. 3.42   Immature fruit of *Ekrixanthera hispaniolae*

Fig. 3.43 Caterpillar
feeding on Ekrixanthera
hispaniolae

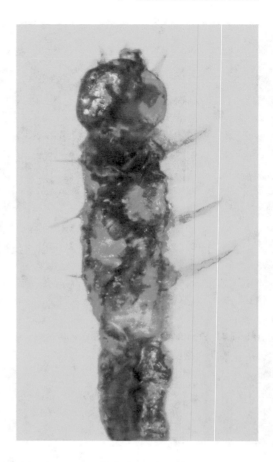

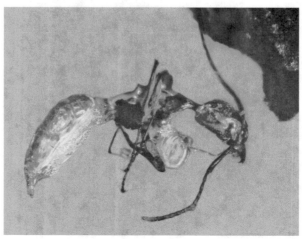

Fig. 3.44 Dolichoderinae ant adjacent to *Ekrixanthera hispaniolae* flower

Fig. 3.45 *Discoflorus neotropicus* flower with adjacent termite

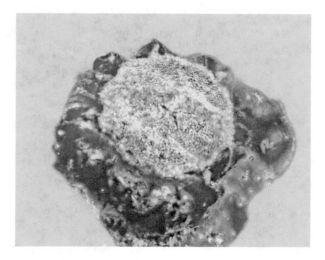

Fig. 3.46 Flattened stigmatic disk of *Discoflorus neotropicus*

Fig. 3.47 Termite pollinator adjacent to *Discoflorus neotropicus*

Fig. 3.48 Two pollinaria of *Discoflorus neotropicus* attached to termite clypeus

Fig. 3.49 Petals of *Cylindrocites browni*

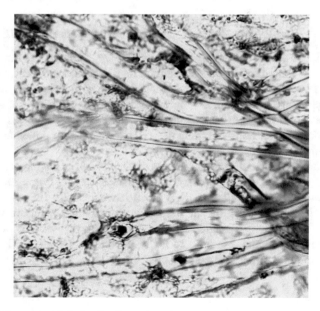

Fig. 3.50 Trichomes on *Cylindrocites browni* petals

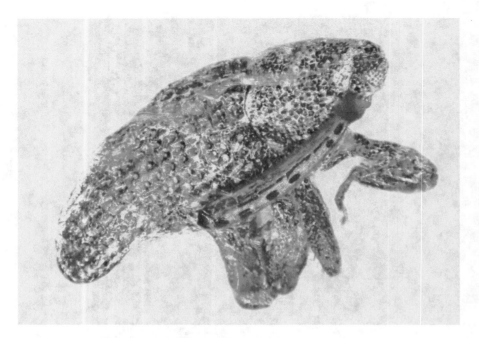

Fig. 3.51 Weevil with attached pollinarium of *Cylindrocites browni*

Based on the features of the West Indian mahogany, the flower of *Swietenia dominicensis* could have come from a medium sized tree that reached 120 feet in height, had leaves with up to 8 leaflets, flowers borne in clusters, and woody fruits with maple-like winged seeds. The flowers of West Indian mahogany are pollinated by bees, moths and thrips.

Treptostemon domingensis, description based on Chambers et al. 2012

This male flower of *Treptostemon domingensis*, which is 4.0 mm in longest diameter, looks something like a starfish with its 6 extended tepals in 2 whorls of 3 (Fig. 3.64). The 9 stamen clusters in whorls of 3 (Fig. 3.65) have short filaments and slightly puberulent anthers with 2–4 pores that open by apical valves. Some anthers appear to be sterile but there is no central cluster of staminodes or pistillodes. The tiny valves on the round fuzzy anthers that open to release the pollen are characteristic of members of the Lauraceae.

Present day species with similar features of *Treptostemon domingensis* are members of the Sweetwood genus *Ocotea*. These are wide ranging tropical and subtropical shrubs and trees with simple, dark green leaves. Some, like the Ecuadorian species, *O. benthamiana*, grow in mountainous habitats. Sweetwoods produce small, edible berries with a single seed that are distributed by birds. Some species produce essential oils that are used to flavor beverages. Others are harvested for timber that contains anti-fungal compounds.

Fig. 3.52 Pollinarium of
Cylindrocites browni

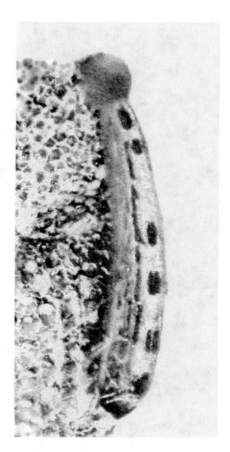

Fig. 3.53 Stingless bee with *Globosites apicola*

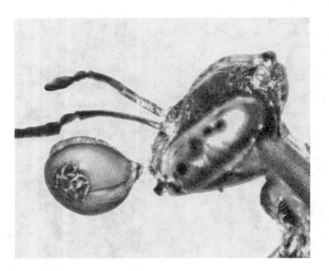

Fig. 3.54 Polliniarium of *Globosites apicola*

Fig. 3.55 *Rudiculites dominicana* pollinarium on head of stingless bee

Fig. 3.56 Bowl of pollinarium of *Rudiculites dominicana*

Various insects are associated with Sweetwood trees, including ants that live in the hollowed out stems and caterpillars of several moths. Pollinators include bees, beetles, flies, moths and butterflies. While assigned to the family Lauraceae, based on features of the stamens, *Treptostemon domingensis* could not be placed in any modern genus.

Klaprothiopsis dyscrita, description based on Poinar et al. 2015

Several fragrant flowers of *Klaprothiopsis dyscrita*, togther with an associated ant and fungus gnat, ended up in a single piece of Dominican amber. The four delicate, membranous petals of *Klaprothiopsis dyscrita* arising above the cuplike ring of short sepals are quite elegant and considering how fragile they are, it is amazing that most still remain attached (Fig. 3.66). While not all 8 stamens remain on all the flowers, those still existing have long, almost straight filaments. Curiously, there are two small sterile stamens (staminodia) with feathery tips positioned at the base of each petal (Fig. 3.67). The ovary is hidden from view but the straight style has a slightly thickened tip.

Fig. 3.57 *Mycophoris elongatus* seed

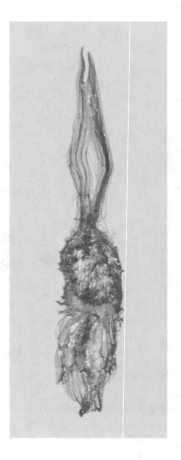

These flowers, which are just under 5 mm in length, are assigned to the stick-leaf family Loasaceae but cannot be placed in any extant genus. They do share some characters with flowers of the genus *Klaprothia* that are found in moist stream banks and well shaded forests from Mexico to Columbia and in Haiti. They occur as herbs or trailing and climbing plants, sometimes using other vegetations for support. The flowers of *Klaprothia* have sepals and petals in sets of 4 with the petals of some species within the size range of those of *Klaprothiopsis dyscrita*. The capsular fruit of *Klaprothiopsis dyscrita* is only 3.0–10.0 mm long and lacks valves so does not split open to release the 2–4 small seeds.

The small lobed terminal staminoids of *Klaprothiopsis dyscrita* were probably producing nectar, which is a feature of members of the Loasaceae. This would explain the presence of the ant (Fig. 3.68) and fungus gnat (Fig. 3.69) adjacent to the flowers of *Klaprothiopsis dyscrita*.

Fig. 3.58 *Synaptomitus orchiphilus* mycelium in embryonic cells of *Mycophoris elongatus*

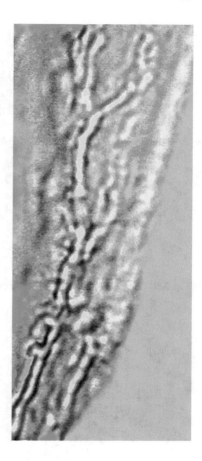

Fig. 3.59 Female *Paleoeuglossa melissiflora*

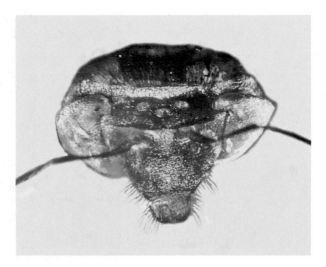

Fig. 3.60 Head of female *Paleoeuglossa melissiflora*

Fig. 3.61 Male Dominican amber orchid bee with possible pollinaria attached

Fig. 3.62   Flower of *Swietenia dominicensis*

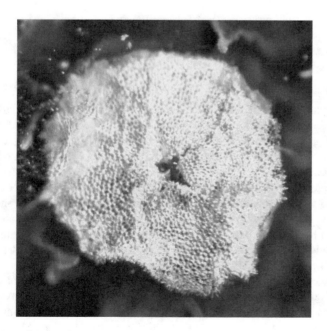

Fig. 3.63   Stigmatic head of *Swietenia dominicensis*

Fig. 3.64 Flower of *Treptostemon domingensis*

Comopellis presbya, description based on Chambers and Poinar 2015

The bisexual flower of *Comopellis presbya* is just under 5 mm in width. It has 5 large spreading sepals, (Fig. 3.70) and 5 small tubular-shaped petals that enclose the stamens (Fig. 3.71). Perhaps this developmental pattern of having the stamens inside cylindrical petals is to attract specific pollinators or to protect the anthers from insect herbivores.

The globular ovary is relatively large, occupying most of the center of the flower. While the ovary is capped by a broad stigma, the style is either very short or non-existent. Also unique to this new genus of the family Rhamnaceae are 5 bilobed glandular appendages equally positioned around the ovary and alternating with the tubular petals. These appendages were probably secreting nectar but the types of pollinators that would be attracted are unknown.

A juvenile nematode is associated with the glandular appendages of *Comopellis presbya* (Fig. 3.72). Even today, it is unusual to find nematodes associated with flowers. Based on its narrow head, small stylet, slender body and pointed tail, it is probably a seed and leaf gall nematode of the genus *Anguina*. Members of this genus are known to invade the developing ovaries or form galls on present day floral parts. This nematode could have been developing in the ovary or in a floral gall of *Comopellis presbya*. Juvenile stages of anguinid nematodes are quite resistance to desiccation. No modern genus of Rhamnaceae could be found that has the unique features of *Comopellis presbya*.

Fig. 3.65   Stamen clusters
of *Treptostemon
domingensis*

Fig. 3.66   Flower of *Klaprothiopsis dyscrita*

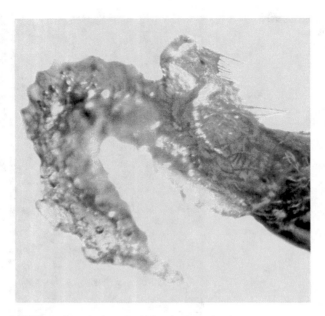

Fig. 3.67 Petal of *Klaprothiopsis dyscrita* with paired staminodes

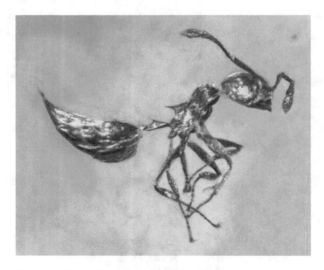

Fig. 3.68 Ant adjacent to *Klaprothiopsis dyscrita* flower

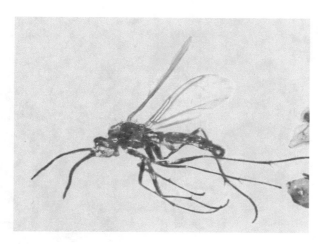

Fig. 3.69   Fungus gnat adjacent to *Klaprothiopsis dyscrita* flower

Fig. 3.70   Flower of *Comopellis presbya*

Fig. 3.71 Stamen
enclosed by petal of
Comopellis presbya

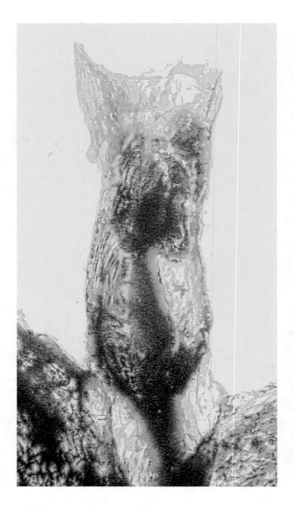

Lobocyclas anomala, description based on Chambers and Poinar 2016a

The slightly over 3 mm wide bisexual flower of *Lobocyclas anomala* has a broad, centrally located, 10-lobed skirt-like disc that supports 3 stamens with kidney shaped anthers (Fig. 3.73). These stamens surround a central short, clubbed style with an expanded stigma (Fig. 3.74). The depressed 3-lobed ovary probably would have matured into a dry, trilobed capsule that is characteristic of fruits of other members of the subfamily Hippocrateoideae in the bittersweet family, Celastraceae. Present day members of this subfamily are mostly climbing vines with simple leaves and flowers that produce dry fruits.

The large discs typical of this subfamily secrete nectar that attracts pollinating bees, flies and beetles. American species of Hippocrateoideae occur throughout Mexico and Central America in lowland tropical moist forests, as well as in other forest types. Inhabitants of the Caribbean islands extracted insecticidal and

Fig. 3.72 Associated nematode adjacent to glandular appendage of *Comopellis presbya*

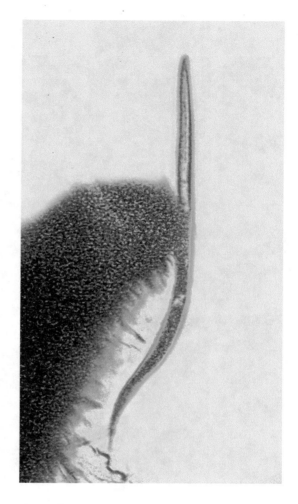

therapeutic compounds from members of the "medicine vine" genus *Hippocratea*, whose flowers closely resemble those of *Lobocyclas anomala*.

3.4 Poaceae

Alarista succina, description based on Poinar and Columbus 2012

It is quite likely that this little 3.0 mm long *Dolichoderus* worker ant was carrying the 13 mm long *Alarista succina* grass spikelet back to its nest when it fell in a drop of resin (Fig. 3.75). Current species of this ant genus are known to transport seeds to their colonies (Fig. 3.76).

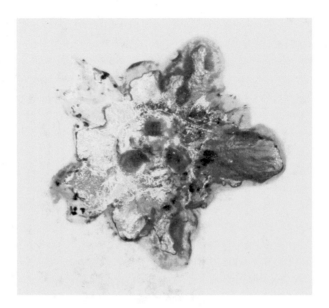

Fig. 3.73 Flower of *Lobocyclas anomala*

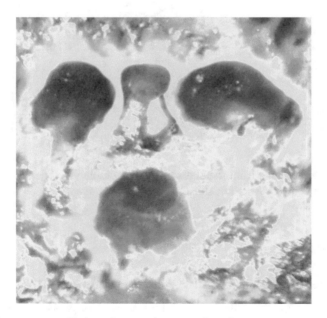

Fig. 3.74 Three stamens of *Lobocycla anomala* surrounding stigma

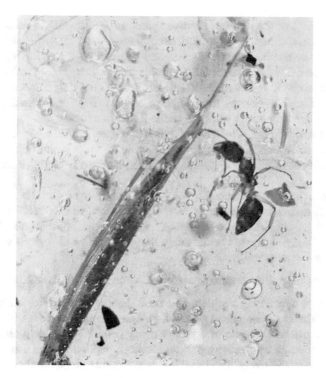

Fig. 3.75 Spikelet of *Alarista succina*

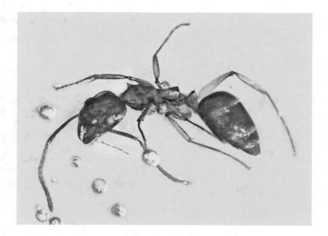

Fig. 3.76 Dolichoderus ant adjacent to *Alarista succina* spikelet

Many features of the *Alarista succina* spikelet align it with the bamboos, one of the most successful and widely distributed grass groups known today. The fossil spikelet share some characters with members of the modern woody bamboo genus *Arthrostylidium*. This genus contains climbing woody bamboos and ranges from Mexico and the West Indies to central Brazil. These climbing bamboos can ascend to the tops of various canopy trees, then pass from tree to tree, ending up on the ground when the supporting trees fall to the forest floor.

Today bamboo shoots, which can be eaten as a green vegetable, are a rich source of vitamins and minerals. Both antimicrobial, anthelmintic and other compounds used in treatment of human ailments have been isolated from the bark and leaves of bamboo.

This is not the only evidence in Dominican amber that ants carried grass spikelets. A grass spikelet of the genus *Panicum* appears to have been dropped by a worker harvest ant (*Pogonomyrmex* sp.) that is now adjacent to the spikelet (Fig. 3.77). *Panicum* grasses occur today thoughout tropical regions of the world, including the Greater Antilles.

Ticodendron palaios, description based on Chambers and Poinar 2014b

This strange 6 mm long pistillate flower of *Ticodendron palaios* is characterized by a spiny peduncle arising from a basal bud. The spiny peduncle leads up to 4 large bracts that surround the base of a bristly ovary (Fig. 3.78). This globular inferior ovary is surmounted by 3 short sepals and 2 short, papillate styles that are flattened against the top of the ovary. The basal bud is surrounded by a number of scales and the inflorescence arises directly from the bud (Fig. 3.79).

The flower is assigned to the present day genus *Ticodendron* of the family Ticodendraceae. This genus was recently described for a single tree species (*T. incognitum*) that occurs in moist, evergreen forests throughout Mesoamerica.

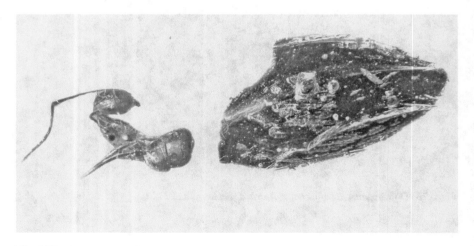

Fig. 3.77 Harvester ant with a *Panicum* grass spikelet

Fig. 3.78 *Ticodendron palaios* flower

This tree reaches 75 feet in height, has fairly broad leaves with sharply pointed tips, and onion bulb-shaped hard fruits; traits that were probably similar to those of *Ticodendron palaios*. The male and female flowers of the extant species of *Ticodendron* occur on separate trees (dioecious) and are wind pollinated, which explains the absence of nectar discs on the fossil.

Protium callianthum, description based on Chambers and Poinar 2013

These two flowers of *Protium callianthum*, both about 6 mm wide, probably fell from the same tree when they were attacked by one or more insect herbivores. Tips of some of the anthers have been nibbled away on one flower and portions of the petals removed on the second flower. Damage is also evident to accompanying leaves. The cupular calyx with basally connate triangular sepals are dwarfed by the 5 recurved darkly blotched petals that possess papillate margins (Fig. 3.80). The 10

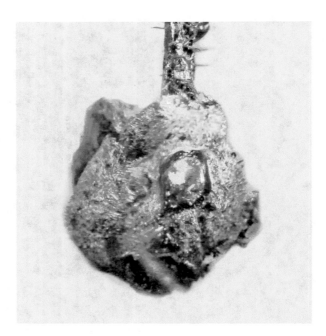

Fig. 3.79   Basal bud of *Ticodendron palaios* with surrounding scales

Fig. 3.80   Flowers of *Protium callianthum*

strongly incurved stamens of 2 different lengths bear linear, laterally dehiscent anthers (Fig. 3.81). The ovoid, glabrous, 5-lobed ovary has a stout 10-lobed stigma.

Flowers of *Protium* are sometimes cryptically dioecious, when apparently perfect flowers with both stamens and an ovary either lack ovules and are functionally male, or have vestigial anthers lacking pollen and are functionally female. The flowers of *Protium callianthum* appear to be functionally male even though they seem to possess functional ovaries.

There are some 170 extant species of *Protium* of the family Burseraceae in the Neotropics today and species of thus genus also occur in Asia and Madagascar. Most species are medium sized trees that grow in wet, shady tropical forests, which is how the Dominican amber forest has been characterized (Poinar and Poinar 1999)

One species of *Protium* found in Mexico and Central America is called the long-leaf copal tree. It produces a resin known as "copal" that is burned as incense and used to make varnish. The fruits of this copal tree are small, red, smooth berries. Other species of *Protium* are used for timber, fruits, firewood, medicinal properties and essential oils that are extracted from leaves, fruits and resin.

Catalpa hispaniolae, description based on Poinar 2016a

It would have been easy to identify this small catalpa tree in the Dominican amber forest since its expanded 13 mm long flowers would have stood out against the broad, heart-shaped leaves. The two bisexual *Catalpa hispaniolae* flowers in separate pieces of Dominican amber are shaped like small pockets since their 5 petals are united (sympetalous) and form a two-lipped tube with the sexual organs located inside, hidden from view (Fig. 3.82). The two stamens with their flattened

Fig. 3.81 Curved stamens of *Protium callianthum*

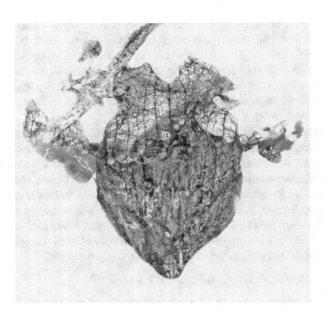

Fig. 3.82 Flower of *Catalpa hispaniolae*

filaments and oblong anthers are inserted at the base of the corolla tube. They can be detected as silhouettes against a bright light positioned behind the flower. The ovary is sessile and possesses a single filiform style.

In a separate piece of Dominican amber was what is considered a catalpa seed (Fig. 3.83). It has the shape and terminal hair tufts characteristic of today's catalpa seeds. These winged seeds are released from long slender pods that begin to develop even before the flowers have fallen. These seeds would have floated down to the forest floor. Catalpa leaves are quite broad and provide protection to hikers when caught in a sudden rainstorm.

There is evidence of insect damage on both of the amber flowers. Present day catalpas are the host plant of sphinx moth caterpillars however the damage could have been made by a number of insects. Occurring throughout North America and East Asia besides the Caribbean region, catalpa wood is used in furniture and cabinetry making and is said to be especially good for constructing acoustic musical instruments.

Trichilia glaesaria, description based on Chambers et al. 2011a

Trichilia glaesaria is a pistillate flower 4.5 mm in length, with a cup-shaped calyx of 5 sepals with broadly acute lobes, 5 large petals and 10 sterile stamens with the filaments forming a cup- shaped tube that is wider than long. The flower has a centrally located ovary with a fairly long style and terminal stigma (Fig. 3.84).

Fig. 3.83 *Catalpa* seed in
Dominican amber

It is curious why the well-formed anthers of *Trichilia glaesaria* would be sterile, even though unisexual flowers are typical of *Trichilia*. The amber tree that furnished these flowers was probably similar to present day *Trichilia* trees that occur in tropical South America and Africa, with spreading crowns and glossy dark leaves. The fruits are roundish capsules that break open to release oval colored seeds. Pollinators of the sweetly scented flowers are bees, butterflies and birds. In Africa, baboons eat and distribute the seeds.

Due to their reddish-brown wood, similar in color to that of true mahogany trees (*Swietenia mahagoni*), *Trichilia* trees are also called "mahogany". The forest mahogany tree in South Africa, (*Trichilia dregeana*), provides wood for construction, carvings, furniture and household implements. Concoctions from *Trichilia* trees are also used in folk medicine as a malaria depressant.

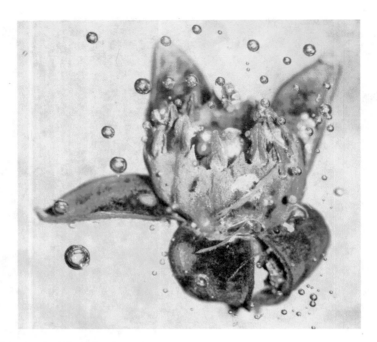

Fig. 3.84 Flower of *Trichilia glaesaria*

Trichilia antiqua, description based on Chambers et al. 2011a

The flower of *Trichilia antiqua*, 3.4 mm in length, also has a cup-shaped calyx of 5 sepals united at the base, but without lobes. There are 6 large petals (1 missing) and 10 sterile stamens (9 with missing anthers). However in contrast to *Trichilia glaesaria* that has a filament tube wider that long, the filaments of *Trichilia antiqua* form a high cup- shaped tube that is distinctly longer than wide (Fig. 3.85).

Both of these amber fossils, as well as three additional flowers of *Trichilia* in separate pieces of Dominican amber, indicate that members of this genus were fairly common in the Dominican amber forest. These additional flowers showed further modifications of the sexual organs, such as reduced male structures (staminodes) with rudimentary filaments and knob-like nonfunctional anthers in functionally female flowers.

Prioria dominicana, description based on Poinar and Chambers 2015b

A number of bisexual *Prioria dominicana* flowers occur in Dominican amber (Fig. 3.86), perhaps because they were growing in pure stands near the resin-producing *Hymenaea* trees. A member of the legume family (Fabaceae), most *Prioria dominicana* flowers are just under 4 millimeters in width. The 2 basal bracts, 5 free sepals, absence of petals, 10 stamens and ovoid ovary with a linear style characterize the flowers (Fig. 3.87).

Fig. 3.85 Flower of *Trichilia antiqua*

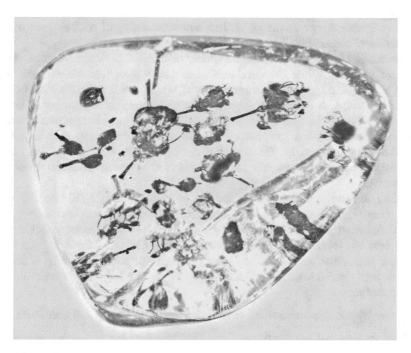

Fig. 3.86 Cluster of *Prioria dominicana* flowers

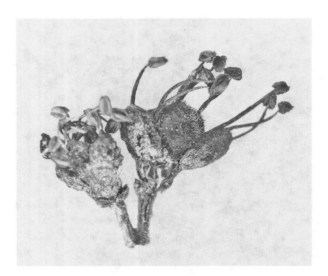

Fig. 3.87 Two flowers of *Prioria dominicana*

The stamen filaments of *Prioria dominicana* are quite long, thus completely exposing the anthers to visiting pollinators. A hemipteran with a long, narrow beak is exploring the base of a *Prioria dominicana* flower (Fig. 3.88). Gall midges (Cecidomyiidae) are important pollinators and some of the amber specimens show developing gall midge larvae as well as adjacent adult gall midges with buds of *Prioria dominicana* (Fig. 3.89).

Present day species of *Prioria* are widely distributed in South America, Africa, Asia and Oceania. While *Prioria dominicana* shares some features with the single living Neotropical species (*Prioria copaifera*), which incidentally does not occur in Hispaniola today, it is actually more similar to *Prioria* trees growing in tropical forests of West Africa. This situation also occurs with *Hymenaea protera* that forms Dominican amber and is morphologically closer to *Hymenaea* trees in East Africa than those in the Americas. These similarities between flowers in Caribbean amber and Africa provides evidence of an ancient connection between these two land masses.

Based on extant species, *Prioria dominicana* was probably a tall tree with shiny, dark, green leaves, small fragrant cream to yellowish flowers and woody, one-seeded flattened pods. It would have grown in rich, loamy, marshy soil close to rivers near sea level. Today *Prioria* trees are often harvested for construction and furniture making. A resin collected from gashes in the trunk and limbs is used for starting fires, producing light and treating wounds. The large seeds are edible and sold in markets.

Salpinganthium hispaniolanum, description based on Poinar and Chambers 2021

Another bisexual legume flower from Dominican amber is *Salpinganthium hispaniolanum*. While slightly larger than *Prioria dominicana* and reaching almost

Fig. 3.88 Hemipteran on *Prioria dominicana* flower

Fig. 3.89 Larva gall midge (left) and adjacent gall midge adult (right) with bud of *Prioria dominicana*

6 mm in width, it is not at all as common as *Prioria* flowers. *Salpinganthium hispaniolanum* has a pair of bracts just below the flower and then 5 quite large sepals and 5 extended, flexible petals. There are 10 stamens with long filaments and a slender, hirsute ovary with a slender style (Fig. 3.90). The ovary is already somewhat elongate and we can assume the fruit would have been a slender pod.

Features of *Salpinganthium hispaniolanum* were probably similar to purpleheart trees of the legume genus *Peltogyne*. Purpleheart trees are medium to large with alternate, compound leaves. Their name comes from the strong, durable wood that has a purplish color and is used for musical instruments, cabinetry and furniture. Purpleheart groves grow along rivers in tropical rain forests from Central to South America, with most species found in the Amazon basin. The small flowers can have white, yellowish or pink petals.

While *Peltogyne* trees still survive today, *Salpinganthium* trees have disappeared. We can only suppose that they succumbed to some unique biological and/or physical events during the past 20–30 million years that decimated their populations.

Fig. 3.90 Flower of *Salpinganthium hispaniolanum*

Persea avita, description based on Chambers et al. 2011b

The bisexual flower of *Persea avita*, with a length of 5.3 mm, had already flowered and lost some of its sepals and stamens before it fell. This damage was probably due to the action of herbivorous insects. There were originally 6 tepals (2 missing), 9 fertile stamens (3 missing) in 3 whorls with glands on the 3rd whorl. The anthers were 4-locular and opened by valves (Fig. 3.91). These and other remaining features, such as the length and shape of the filaments and anthers, open valves on the anthers and a series of glands associated with the stamens place *Persea avita* in the family Lauraceae.

Today's *Persea* spp. are medium sized trees with simple, smooth leaves. They occur in Central and South America, Mexico, the Caribbean, Africa and southeast Asia. The most well-known member of the genus is the avocado (*Persea americana*), which is thought to have originated in Mexico and then spread to other areas. Few realize that the avocado is simply a large fleshy berry with a sizable seed inside.

The flowers of present day *Persea* trees are greenish yellow to white and very small and inconspicuous, similar in size to those of the fossil. Scale insects (Hemiptera: Coccidae) are one of the most serious pests of avocado today. As these insects drain the tree of sap, they secrete large amounts of honeydew. Various black

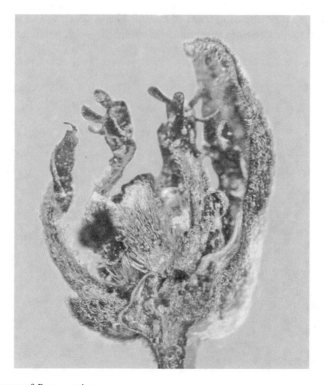

Fig. 3.91 Flower of *Persea avita*

sooty molds, especially those of the genus *Alternaria*, develop on this honeydew and can slowly weaken and destroy the plants. The fungal spore attached to one of the stamen filaments of *Persea avita* closely resembles spores of *Alternaria* and could indicate the presence of a disease complex (Fig. 3.92)

Virola dominicana, description based on Poinar and Steeves 2013

Based on the habits of present day *Virola* trees of the nutmeg family (Myristicaceae), *Virola dominicana* would have been a medium size tree with large elliptical leaves with pointed tips. The flowers would have been tiny, reddish-brown and borne on stalks attached to the tree limbs. Today *Virola* trees are absent from Hispaniola but occur in Central and South American lowland and cloud rainforests.

The abundance of *Virola dominicana* in the amber forest is indicated by some 24 male flowers encased in different pieces of Dominican amber. These brown to orange-brown flowers are about 5 mm in diameter and normally possess 3 (occasionally 2) tepals that are covered with short, stubby hairs (Fig. 3.93). In the center of the flowers are 3 stamens with partly fused filaments but separate anthers shedding boat-shaped pollen.

Variation in tepal number occurs in current species of *Virola* with some having only 2 and others 4 tepals. While some flowers of *Virola dominicana* only had 2 tepals, none had four. The absence of female flowers of *Virola dominicana* in amber can be explained by examining traits of current *Virola* species. The female flowers, which occur on separate trees from male flowers, have thick pedicels and rarely fall

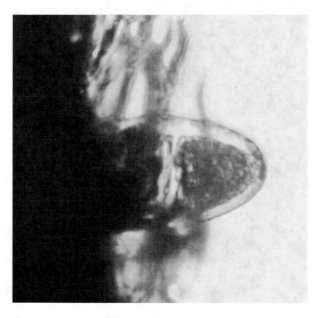

Fig. 3.92 Fungal spore on stamen of *Persea avita*

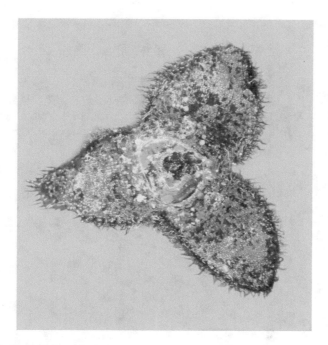

Fig. 3.93 *Virola dominicana* flower

to the ground, while male flowers, which have thin pedicels, fall after they have shed their pollen.

Among the flowers of *Virola dominicana* are leafhoppers (Hemiptera: Cicadellidae) (Fig. 3.94) and worker *Cephalotes* ants (Hymenoptera: Formicidae) (Fig. 3.95). These insects, some of which could have served as pollinators, were undoubtedly attracted to nectar secreted from the tips of the short, thick hairs lining the surface of the tepals.

Cephalotes ants have several amazing features. If they fall from their arboreal nests, they are able to "glide" in the air back to the tree surface. Then, the soldier ants, as well as some large workers like the one taking nectar from *Virola dominicana*, have very wide, armored heads. They station themselves at the entrance of the small hole leading to the colony's nest. At this post, they inspect all ants entering the colony, readily admitting returning nest mates, but blocking the entrance with their large heads when intruders approach.

The bright orange-red fruits of present day *Virola* trees are considered to be highly nutritious and are especially prized by birds and spider monkeys. Psychedelic drugs have been isolated from the red seeds that develop inside the roundish green fruits. In the past, these drugs were widely used to induce mystical visions and hallucinations.

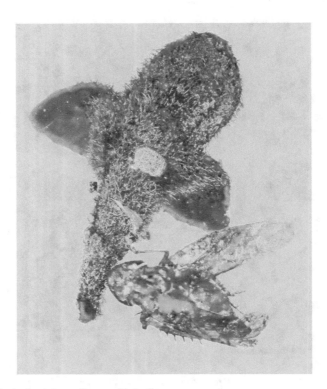

Fig. 3.94 *Virola dominicana* flower with leafhopper

Strychnos electri, description based on Poinar and Struwe 2016

Finding two *Strychnos electri* flowers in Dominican amber was quite a surprise since living trees of this interesting genus of the family Loganiaceae have never been reported from the Dominican Republic. Other members of the genus *Strychnos* occur in tropical Asia, Africa, India, and South America, with a single species in Cuba.

Strychnos electri flowers are bisexual and the 5 petals form a long, trumpet-shaped floral tube that is nearly 5 mm in length and covered with tomentose hairs (Fig. 3.96). There are 5 stamens but only the tips of the anthers are visible rising above the petal lobes (Figs. 3.97 and 3.98). Spherical pollen grains adjacent to the anthers show that the flower was in anthesis when it fell in the resin (Fig. 3.99). While the ovary, along with the basal sepals, are missing, the single style with its capitate stigma is quite long, extending upward more than half the length of the floral tube (Fig. 3.96).

Butterflies are known to pollinate current *Strychnos* flowers, probably because their long proboscises can reach to the bottom of the extended floral tubes. One possible contender in Dominican amber would be a metalmark butterfly of the genus *Napaea* (Lepidoptera: Riodinidae) (Poinar and Poinar 1999) (Fig. 3.100).

Fig. 3.95 *Cephalotus* ant collecting nectar from *Virola dominicana* flower

Present day *Strychnos* species are herbs, shrubs, trees and lianas. However features of *Strychnos electri* align it with the tropical South America strychnine vine (*Strychnos toxifera*) from which strychnine poison (curare) is extracted from the bark and stems. The poison is applied to the tips of blowgun darts made from the spines of the mingucha palm and "cotton" from the seeds of the *Bombax* cotton tree. Using a blowgun made from hallowed out stems of another palm, the poisoned darts were then blown at prey. The poison paralyzed the victim's respiratory muscles. These blowguns were used in place of guns since the hunter could sneak up on the prey and shoot them one by one without making any noise that would scatter the remainder. Since curare is harmless when taken orally, it posed no problem for those eating prey killed in this manner.

Fig. 3.96 Flower of
Strychnos electri

By knowing the location of these vines in dense tropical forests and how to extract the poison, various tribes could control their collection, use and trade. Curare had its beneficial use in the past as a treatment of tetanus and a narcotic for patients during surgical operations.

Licania dominicensis, descriptions based on Chambers and Poinar 2010 and Poinar et al. 2008a

Theses curious, small, bisexual flowers that appear in 5 separate pieces of Dominican amber were originally described as *Lasiambix dominicensis* and tentatively placed in the family Fabaceae (Poinar et al. 2008a). They were later reassigned to the genus *Licania* in the family Chrysobalanaceae (Chambers and Poinar 2010).

Fig. 3.97 Top of *Strychnos electri* flower

Fig. 3.98 Petals of *Strychnos electri*

Fig. 3.99 Pollen grains of
Strychnos electri

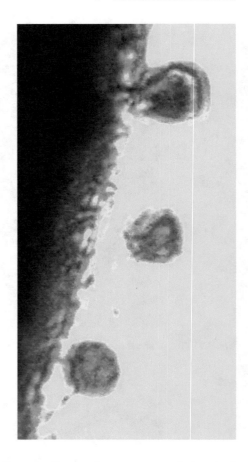

The five short, stubby sepals (3.6–3.7 mm in diameter) appear as petals but the 5 actual petals have fallen (early deciduous). The 3 dark stamens, with their short, incurved filaments and oval anthers, are positioned opposite 3 of the sepals (Fig. 3.101). The small style is located between the two sepals that lack stamens. The styles fall off as the small, oval fruits, with their single seed, mature and many can be found free in amber (Fig. 3.102).

Based on features of present day *Licania* trees, we can propose that the Dominican amber *Licania dominicensis* was a medium sized tree with elliptical dark green leaves and flowers borne on long pendent stalks emerging from the branch tips. The dominant flower clusters are why some *Licania* species are called "tassel" trees. These trees are often planted in gardens or along streets. The bark and leaves of some species are used in folk medicine. The seeds are high in oils and are used for making candles and soap. The wood is used in rural construction and for fence posts.

While there are no representatives of *Licania* in the entire Greater Antilles today, *Licania dominicensis* shows that the genus was present in the Caribbean some 20–30 million years ago.

Fig. 3.100 A metalmark butterfly of the genus *Napaea* in Dominican amber

Fig. 3.101 *Licania dominicensis* flower

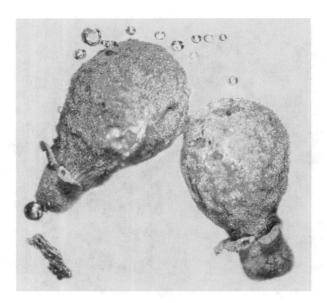

Fig. 3.102   *Licania dominicensis* fruits

***Pharus primuncinatus*,** description based on Poinar and Judziewicz 2005

Various plants have seed pods or seeds that bear hooks for attachment to the pelage of mammals or feathers of birds. This condition, dispersal by attachment to animals, is called epozoochory. The 10 mm long grass flower of *Pharus primuncinatus* in Dominican amber is triangular in shape and surrounded by mammalian hairs (Fig. 3.103). Along the edge of one of its floral elements (lemmas) are numerous hooks that clasp a strand of hair (Fig. 3.104). These hooked hairs are modified to attach themselves to animals.

An analysis of the hairs associated with *Pharus primuncinatus* shows them to be from a carnivorous mammal. Based on a list of various mammals that carry similar spikelets today, it was considered most likely that the hairs came from a jaguar-like feline (Poinar and Poinar 1999). In this case the animal carrying the *Pharus primuncinatus* spikelet probably tried to remove it by brushing up against a *Hymenaea protera* tree, thus leaving the irritant stuck to some resin.

Today, there are some seven species of Neotropical broad-leaved, herbaceous bamboos of the genus *Pharus*, all of which bear hooked hairs for animal dispersal. This kind of transport accounts for the isolated presence of terrestrial plants in unique habitats, such as on islands when wind dispersal would not have been possible. Animals, especially birds, are known to distribute plants in this manner over hundreds of miles. The associated fungus gnat (Diptera: Mycetophilidae) (Fig. 3.103) may have been breeding in soil among the roots of *Pharus primuncinatus* since that association occurs today.

Fig. 3.103 *Pharus*
primuncinatus spikelet
with fungus gnat

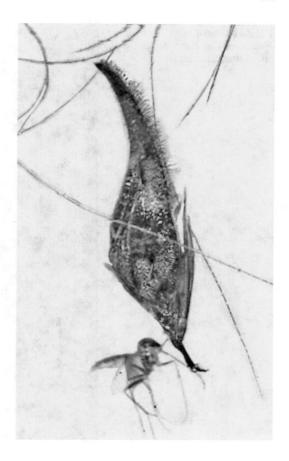

3.5   Arecaceae (Palms)

Based on palm flowers in amber, both fan and pinnate leaved palms flourished in the Dominican amber forest, just as they occur today throughout the Neotropics.

Palaeoraphe dominicana, description based on Poinar 2002

Palaeoraphe dominicana is one of the larger flowers in Dominican amber, measuring just over 11 mm in diameter. It is dominated by 3 broad outstretches petals. The 3 sepals form a tube below the petals and are only noticeable in side view. This striking flower was in full bloom when it fell into the resin and three of the 6 stamens are quite noticeable as they protrude from the sides of the flower between the petals (Fig. 3.105). The other 3 stamens rest on the surface of the petals (Fig. 3.106). The strongly ridged ovary is located in the center of the flower.

Fig. 3.104 Hooks on
Pharus primuncinatus with
attached mammalian hair

Palaeoraphe dominicana could not be placed in any current genus. Palms with similar flowers are hesper fan palms (*Brahea* spp.) that occur from Mexico through Central America.

However, neither *Palaeoraphe dominicana* or hesper palms occur in Hispaniola today. Whether palm feeding bruchid beetles, such as the Dominican amber *Caryobruchus dominicanus* (Poinar 1999) (Fig. 3.107) were responsible for their demise is unknown. Female palm beetles, which occur in Dominican amber, deposit eggs on developing palm fruits and the larvae develop in the seeds. After emerging from the empty seeds, the adult beetles consume palm pollen.

Trithrinax dominicana, description based on Poinar 2002

The delicate, bisexual palm flower of *Trithrinax dominicana* with its 3 broad sepals and 3 extended petals is just over 3 mm long (Fig. 3.108). The 6 erect stamens bare roundish anthers with their tips bent inwards and the 3-lobed ovary possess

Fig. 3.105   *Palaeoraphe dominicana* flower

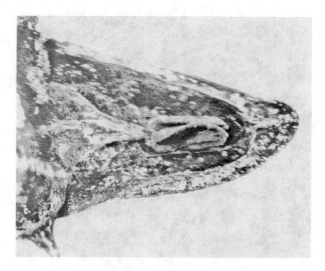

Fig. 3.106   Stamen of *Palaeoraphe dominicana*

Fig. 3.107 Palm beetle in Dominican amber

Fig. 3.108 *Trithrinax dominicana* flower

3 min styles. Based on current species of *Trithrinax* fan palms, the flower of *Trithrinax dominicana* came from a large inflorescence borne near the crown of the plant. The palm trunk would have been quite spiny since these palms keep their dead leaves as a thick protective coat. The sturdy leaves of these palms, whose fibers are quite resistant, are used today in making handicrafts and primitive clothing. The seeds contain useful oils and the fruits are used to produce fermented beverages.

Species of *Trithrinax* are found today in forest clearings in South America but are absent from the Greater Antilles. Its disappearance could have been due to Pliocene-Pleistocene cooling events, recent habitat disturbances or predation. The present day descendants of a small, slender weevil in Dominican amber develop on palms and it is possible that *Bicalcasura maculata* (Poinar and Legalov 2013) (Fig. 3.109) used *Trithrinax dominicana* as its host plant. The small phoretic stages of uropodid turtle mites are attached to the legs of *Bicalcasura maculata*.

Fig. 3.109 Palm weevil (*Bicalcasura maculata*)

Fig. 3.110   Mature pistillate flower of *Roystonea palaea*

Roystonea palaea, description based on Poinar 2002

The mature pistillate flower of *Roystonea palaea* is slightly under 2 mm in length and shows evidence of insect damage on the developing ovary. Three imbricate sepals and 3 ovate petals are closely appressed to the sides of the expanding ovary. The 3 curved stigmas at the tip of the developing fruit are curved back in the form of horns (Fig. 3.110). The staminate flower has 3 imbricate sepals, 3 distinct petals and 6 stamens with elongate dorsifixed anthers (Fig. 3.111).

Species of *Roystonea* or "royal" palms are all pinnately veined with unarmed, smooth stems that resemble stone pillars. They can extend up to 140 feet in height and the white flower clusters appear just beneath the tree crowns. The sexes are separate but both are formed on the same plant (monoecious). The fruit is a type of berry (drupe) that turns purple when ripe.

Fig. 3.111   Staminate flower of *Roystonea palaea*

Roystonea is one of the palm lineages still existing in Hispaniola. With their whitish stems and graceful leaves, royal palms today are used as decorative plants. The tender centers of Royal palm stems are used in salads and the seeds have been used as a substitute for coffee.

A palm bug, *Paleodoris lattini* (Poinar and Santiago-Blay 1997), in Dominican amber may have been feeding on *Roystonea palaea* (Fig. 3.112). This insect, barely 3 mm in length, has a unique morphology that is adapted for living between folded pinnate palm leaves like those of the present fossil. The palm bug's body is extremely flat, so much so that it is unrecognizable from the side view (Fig. 3.113). This allows the insect to live between the folds of the palm leaves, where it imbibes sap through its beak. Another modification is that the genital system of the males has swiveled some 45 degrees so they can mate when adjacent to the female. While there are records of palm bugs feeding on Royal palms today in Cuba, native palm bugs appear to have disappeared in Hispaniola.

Fig. 3.112 Palm bug
(*Paleodoris lattini*)

3.6   Indirect Evidence of Plant Families

While there is no direct evidence of the presence of some plant families in Dominican amber based on flowers, their existence can be determined by the presence of insects that are dependant on specific plant genera. One example is the presence of fig wasps (Agaonidae) (Fig. 3.114) in amber, thus showing that trees of the family Moraceae colonized the forest at some point in time. Fig wasps are dependent on fig trees for raising their brood and the fig trees are dependent on the wasps for pollination.

Another example of using indirect evidence is the presence of stalk-winged damselflies, such as *Diceratobasis worki* (Poinar 1996) (Fig. 3.115), that indicates that epiphytic bromeliads of the family Bromeliaceae existed in the amber forest since recent damselflies of this lineage deposit their eggs in water collected in the bowl shaped leaf bases of tank bromeliads. Other life forms entrapped in Dominican amber that are known to inhabit bromeliad leaf bases include tadpoles (Fig. 3.116) (Poinar and Poinar 1999), small frogs of the genus *Eleutherodactylus* (Fig. 3.117)

Fig. 3.113 Side view of
palm bug (*Paleodoris
lattini*)

Fig. 3.114 Fig wasp in Dominican amber

Fig. 3.115 Stalk-winged damselfly in Dominican amber

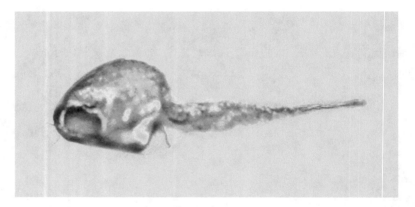

Fig. 3.116 Tadpole in Dominican amber

(Poinar and Cannatella 1987), salamanders such as *Palaeoplethodon hispaniolae* (Poinar and Wake 2015) (Fig. 3.118) and a diving beetle (*Copelatus* sp. Coleoptera: Dytiscidae) (Poinar and Poinar 1999) (Fig. 3.119).

The number of plant families in Dominican amber certainly would be increased if various fruits and seeds could be identified. Many occur with accompanying insects like a small berry with an adjacent feeding stinkbug nymph (Hemiptera: Pentatomidae) (Fig. 3.120). Others are very unique, like a rolled up spiked seed with spurs on the outer surface (Fig. 3.121). But unless they are common today, identification is very difficult.

Fig. 3.117   Frog in Dominican amber

Fig. 3.118   Salamander in Dominican amber

Fig. 3.119 Diving beetle
in Dominican amber

3.7   Summary of Dominican amber Flowers

Based on direct and indirect evidence, representatives of over 20 different plant
families enhanced the Dominican amber forest with their beauty, grace and mysterious qualities. Other pieces of Dominican amber reveal moss, lichen, liverworts and
various fungi. Together they all indicate that the Dominican amber forest was basically a tropical moist forest, similar to those found in Central and South America
today (Poinar and Poinar 1999).

While 20 flowers in Dominican amber can be placed in current families and genera, 5 belong to unknown families, showing that extinctions have occurred at the
species, genus and family levels over the past 20–30 million years (Table 3.1). None
of the Dominican amber flowers represent present day species.

The original Dominican amber forest had a distinct canopy layer composed of
legumes such as *Hymenaea protera*, *Prioria dominicana* and *Salpinganthium hispaniolanum*, with emergent trees like *Swietenia dominicensis* and others extending
beyond the canopy. The subcanopy and understory were represented by royal palms,

Fig. 3.120 Stink bug nymph feeding on berry in Dominican amber

Fig. 3.121 Spiked seed in Dominican amber

Table 3.1 Dominican amber flowers (N = 38 species)

Flower	Systematic placement	References
Alarista succina	Poaceae	Poinar and Columbus (2012)
[a]Fig wasp (Agaonidae)	Moraceae	Poinar and Poinar (1999)
Brevitrimaris arcuatus	Monocot	Chambers and Poinar (2016b)
Campopetala dominicana	Fabaceae	Poinar and Chambers (2016)
Catalpa hispaniolae	Bignoniaceae	Poinar (2016a)
Comopellis presbya	Rhamnaceae	Chambers and Poinar (2015)
Cylindrocites browni	Orchidaceae	Poinar (2016c)
Dasylarynx anomalus	Monocot	Poinar and Chambers (2015a)
[a]*Diceratobasis worki*	Bromeliaceae	Poinar (1996)
Discoflorus neotropicus	Apocynaceae	Poinar (2017c)
Distigouania irregularis	Rhamnaceae	Chambers and Poinar (2014a)
Entada hispaniolae	Fabaceae	Poinar and Chambers (2016)
Ekrixanthera hispaniolae	Urticaceae	Poinar et al. (2016)
Globosites apicola	Orchidaceae	Poinar (2016b)
Hymenaea protera	Fabaceae	Poinar (1991)
Klaprothiopsis dyscrita	Loasaceae	Poinar et al. (2015)
Licania dominicensis	Chrysobalanaceae	Poinar et al. (2008a) and Chambers and Poinar (2010)
Lobocyclas anomala	Celastraceae	Chambers and Poinar (2016a)
Meliorchis caribea	Orchidaceae	Ramirez et al. (2007)
Mycophoris elongatus	Orchidaceae	Poinar (2017a)
Palaeoraphe dominicana	Arecaceae	Poinar (2002)
Persea avita	Lauraceae	Chambers et al. (2011b)
Pharus primuncinatus	Poaceae	Poinar and Judziewicz (2005)
Phyrtandrus pentalepidus	Unknown	Chambers and Poinar (2016c)
Prioria dominicana	Fabaceae	Poinar and Chambers (2015b)
Protium callianthum	Burseraceae	Chambers and Poinar (2013)
Pseudhaplocricus hexandrus	Commelinaceae	Poinar and Chambers (2015c)
Roystonea palaea	Arecaceae	Poinar (2002)
Rudiculites dominicana	Orchidaceae	Poinar (2016b)
Salpinganthium hispaniolanum	Fabaceae	Poinar and Chambers (2021)
Senegalia eocaribbeansis	Fabaceae	Poinar and Chambers (2016)
Strychnos electri	Loganiaceae	Poinar and Struwe (2016)
Swietenia dominicensis	Meliaceae	Chambers and Poinar (2012b)
Ticodendron palaios	Ticodendraceae	Chambers and Poinar (2014b)
Treptostemon domingensis	Lauraceae	Chambers et al. (2012)
Trichilia antiqua	Meliaceae	Chambers et al. (2011a) and Chambers and Poinar (2012a)
Trichilia glaesaria	Meliaceae	Chambers et al. (2011a) and Chambers and Poinar (2012a)

(continued)

Table 3.1 (continued)

Flower	Systematic placement	References
Trithrinax dominicana	Arecaceae	Poinar (2002)
Trochanthera lepidota	Unknown	Poinar et al. (2008b)
Virola dominicana	Myristicaceae	Poinar and Steeves (2013)

[a]Indirect evidence

figs, laurels and mimosoids. The shrub layer included other types of palms as well as acacias. Grasses colonized the forest floor. Orchids, bromeliads, ferns, moss, lichens and vines covered the trees, and various lianas sent their twisting stems over shrubs and between neighboring trees.

Feathers and hairs in Dominican amber show evidence of birds and mammals, including bats. You can almost visualize the small clouds of birds rising up from the flower heads of mimosoids, the disturbance sending down showers of tiny white flowers, and the birds circling around to re-settle on the plants.

Key to Dominican amber Flowers

1. Flowers in spikelets – 2

    1A. Flowers not in spikelets – 3

2. Lemma with hooked hairs – *Pharus primuncinatus*

    2A. Lemma lacking hooks, with long awn – *Alarista succina*

3. Petals absent – 4

    3A. Petals present – 6

4. Flower head composed of many small flowers with 3–4 closely appressed stamens – *Trochanthera lepidota*

    4A. Flowers not as above – 5

5. Flower with 5 stamens subtended by narrow sepals; hairy pistillode in flower center – *Ekrixanthera hispaniolae*

    5A. Flower with 5 merous calyx and 10 stamens – *Prioria dominicana*

6. Flower with 3 large and 2 minute (scale-like) petals – *Hymenaea protera*

    6A. Flowers with petals approximately same size – 7

7. Flowers with 3 or 4 petals (or tepals) – 8

    7A. Flowers with 5 or 6 petals (or tepals) – 17

8. Flower with 3 petals, 3 curved stamens in flower center – *Brevitrimaris arcuatus*

    8A. Flowers not as above – 9

9. Flower with 3 sepals and 6 outstreched stamens – *Pseudhaplocricus hexandrus*

  9A. Flowers not as above – 10

10. Flower with 3 fused petals forming a long floral tube with inner trichomes – *Dasylarynx anomalus*

  10A. Flowers not as above – 11

11. Flower with 3 petals and 3 stamens – *Virola dominicana*

  11A. Flowers not as above – 12

12. Flower with 3 short tepals, 2 short styles and bristly ovary – *Ticodendron palaios*

  12A. Flowers not as above – 13

13. Flower with 3 sepals, 3 petals and 6 stamens (Araceae) – 13

  13A. Flowers not as above – 14

14. Flower unisexual, with 3 outstretched stamens and 3 stamens over petals – *Palaeoraphe dominicana*

  14A. Flowers not as above – 15

15. Flower bisexual, 6 erect stamens, 3 lobed ovary with 3 styles – *Trithrinax dominicana*

  15A. Flowers not as above – 16

16. Flower unisexual with 6 erect stamens; sepals and petals closely appressed to ovary – *Roystonea palaea*

  16A. Flower with 4 petals and 8 stamens – *Klaprothiopsis dyscrita*

17. Flower with 6 extended tepals and 9 stamens – *Treptostemon domingensis*

  17A. Flowers not as above – 18

18. Flower with 3 rings of stamens and staminoids – *Phyrtandrus pentalepidus*

  18A. Flowers not as above – 19

19. Flowers with 10 or more stamens with glands on anthers – 20

  19A. Flowers not as above – 23

20. Petals forming closed tube around 10 stamens; glands on one end of anther – *Entada hispaniolae*

  20A. Stamens number between 20 and 80 – 21

21. Stamens 20–30, anthers with pollen sacs at either end – *Campopetala dominicana*

  21A. Flowers not as above – 22

22. Stamens 50–80 per flower, anthers with stalked apical glands – *Senegalia eocaribbeansis*

    22A. Flowers not as above – 23

23. Sepals larger than petals; cluster of tangled hairs in floral center – *Distigouania irregularis*

    23A. Flowers not as above – 24

24. Flower with 5 tubular petals, each enclosing a stamen – *Comopellis presbya*

    24A. Flowers not as above – 25

25. Flower with 5 petals forming a floral tube with filaments of 10 stamens – *Swietenia dominicensis*

    25A. Flowers not as above – 26

26. Flower with 5 short petals and 10-lobed disc with 3 stamens – *Lobocyclas anomala*

    26A. Flowers not as above – 27

27. Flower sympetalous with a two-lipped tube enclosing 2 stamens – *Catalpa hispaniolae*

    27A. Flowers not as above – 28

28. Flowers with 5 petals with papillate margins and 10 stamens – *Protium callianthum*

    28A. Flowers not as above – 29

29. Flowers with 5 or 6 petals and shallow or deep filament tube – 30

    29A. Flowers not as above – 31

30. Petals 5, filament tube shorter than wide – *Trichilia glaesaria*

    30A. Petals 6, filament tube longer than wide – *Trichilia antiqua*

31. Six tepals, 9 stamens in 3 whorls with glands on the 3rd whorl – *Persea avita*

    31A. Flowers not as above – 32

32. Flowers with 5 sepals, 5 petals and 10 stamens – *Salpinganthium hispaniolanum*

    32A. Flowers not as above – 33

33. Five petals form long floral tube with 5 stamens inside; long style protruding from center of floral tube – *Strychnos electri*

    33A. Flowers not as above – 34

34. Flowers with 5 sepals, 5 petals, 3 dark stamens with short incurved filaments and a small style – *Licania dominicensis*

    34A. Flowers not as above – 35

35. Flowers bisexual with 5 sepals, 5 long petals, 5 stamens and pollen produced in pollinia; attached to termite pollinator – *Discoflorus neotropicus*

    35A. Flowers reproduce by forming pollinia attached to beetles or bees – 36

36. Elongate pollinarium attached to weevil – *Cylindrocites browni*

    36A. Pollinaria attached to stingless bees – 37

37. Pollinarium attached to mesoscutellum of bee – *Meliorchis caribea*

    37A. Pollinaria attached to head of bee – 38

38. Club-shaped pollinarium attached to side of head – *Rudiculites dominicana*

    38A. Spherical pollinarium attached to lip – *Globosites apicola*

References

Chambers KL, Poinar GO Jr (2010) The Dominican amber fossil *Lasiambix* (Fabaceae, Caesalpinioideae?) is a *Licania* (Chrysobalanaceae). J Bot Res Inst Texas 4:219–220

Chambers KL, Poinar GO Jr (2012a) Additional fossils in Dominican amber give evidence of anther abortion in mid-Tertiary *Trichilia* (Meliaceae). J Bot Res Inst Texas 6:561–565

Chambers KL, Poinar GO Jr (2012b) A mid-Tertiary fossil flower of *Swietenia* (Meliaceae) in Dominican amber. J Bot Res Inst Texas 6:123–127

Chambers KL, Poinar GO Jr (2013) A fossil flower of the genus *Protium* (Burseraceae) in mid-Tertiary amber from the Dominican Republic. J Bot Res Inst Texas 7:367–373

Chambers KL, Poinar GO Jr (2014a) *Distigouania irregularis* (Rhamnaceae) gen. et sp. nov. in mid-Tertiary amber from the Dominican Republic. J Bot Res Inst Texas 8:555–561

Chambers KL, Poinar GO Jr (2014b) *Ticodendron palaios* sp. nov. (Ticodendraceae), a mid-Tertiary fossil flower in Dominican amber. J Bot Res Inst Texas 8:563–568

Chambers KL, Poinar GO Jr (2015) *Comopellis presbya* (Rhamnaceae) gen. et sp. nov. in mid-Tertiary amber from the Dominican Republic. J Bot Res Inst Texas 9:361–367

Chambers KL, Poinar GO Jr (2016a) *Lobocyclas anomala*, a new genus and species of Celastraceae subfamily Hippocrateoideae in Dominican amber. J Bot Res Inst Texas 10:137–140

Chambers KL, Poinar GO Jr (2016b) *Brevitrimaris arcuatus* gen. et sp. nov., a monocotyledonous fossil flower from mid-Tertiary amber deposits in the Dominican Republic. J Bot Res Inst Texas 10:141–146

Chambers KL, Poinar GO Jr (2016c) *Phyrtandrus pentalepidus* gen, et sp. nov., a unique fossil flower of unknown affinity found in mid-Tertiary Dominican amber. J Bot Res Inst Texas 10:449–453

Chambers KL, Poinar GO Jr, Brown AE (2011a) Two fossil flowers of *Trichilia* (Meliaceae) in Dominican amber. J Bot Res Inst Texas 5:463–468

Chambers KL, Poinar GO Jr, Brown AE (2011b) A fossil flower of *Persea* (Lauraceae) in Tertiary Dominican amber. J Bot Res Inst Texas 5:457–462

Chambers KL, Poinar GO Jr, Chanderbali AS (2012) *Treptostemon* (Lauraceae), a new genus of fossil flower from mid-Tertiary Dominican amber. J Bot Res Inst Texas 6:551–556

Draper G, Mann P, Lewis JF (1994) Hispaniola. In: Donovan S, Jackson TA (eds) Caribbean geology: an introduction. The University of the West Indies Publishers' Association, Kingston, pp 129–150

Iturralde-Vinent MA, MacPhee RDE (1996) Age and Paleogeographic origin of Dominican amber. Science 273:1850–1852

Iturralde-Vinent MA, MacPhee RDE (2019) Remarks on the age of Dominican amber. Palaeoentomology 2:236–240

Poinar GO Jr (1991) *Hymenaea protera* sp.n. (Leguminosae: Caesalpinoideae) from Dominican amber has African affinities. Experientia 47:1075–1082

Poinar GO Jr (1996) A fossil stalk-winged damselfly, *Diceratobasis worki* spec. nov. from Dominican amber, with possible ovipositional behavior in tank bromeliads (Zygoptera: Coenagrionidae). Odonatologica 25:381–385

Poinar GO Jr (1998) *Paleoeuglossa melissiflora* gen. n., sp. n. (Euglossinae: Apidae), fossil orchid bees in Dominican amber. J Kansas Entomol Soc 71:29–34

Poinar GO Jr (1999) A fossil palm bruchid, *Caryobruchus dominicanus* sp. n. (Pachymerini: Bruchidae) in Dominican amber. Entomol Scand 30:219–224

Poinar GO Jr (2002) Fossil palm flowers in Dominican and Mexican amber. Bot J Linn Soc 138:57–61

Poinar GO Jr (2016a) The first fossil flowers of Bignoniaceae (Lamiales): *Catalpa hispaniolae* sp. nov. in Dominican Republic amber. Novon 25:57–63

Poinar GO Jr (2016b) Orchid pollinaria (Orchidaceae) attached to stingless bees (Hymenoptera: Apidae) in Dominican amber. N Jb Geol Paläont 279:287–293

Poinar GO Jr (2016c) Beetles with orchid pollinaria in Dominican and Mexican amber. Am Entomol 62:180–185

Poinar GO Jr (2017a) Two new genera, *Mycophoris* gen. nov., (Orchidaceae) and *Synaptomitus* gen. nov. (Basidiomycota) based on a fossil seed with developing embryo and associated fungus in Dominican amber. Botany 95:1–8

Poinar GO Jr (2017b) Developmental stages of the fungus, *Synaptomitus orchiphilus*, in the germinating seed, *Mycophoris elongatus* (Orchidaceae), in Dominican amber. Hist Biol 31:1–5

Poinar GO Jr (2017c) Ancient termite pollinators of milkweed flowers in Dominican amber. Am Entomol 63:52–56

Poinar GO Jr (2021) The fossil record of orchids in amber. In: Djordjevic V (ed) Orchidaceae, characteristics, distribution and taxonomy. Nova Science Publishers, New York, pp 1–26. ISBN: 978-1-53614-170-2

Poinar GO Jr, Cannatella DC (1987) An upper Eocene frog from the Dominican Republic and its implication for Caribbean biogeography. Science 237:1215–1216

Poinar GO Jr, Chambers KL (2015a) *Dasylarynx anomalus* gen. et sp. nov., a tubular monocotyledon-like flower in mid-Tertiary Dominican amber. J Bot Res Inst Texas 9:121–128

Poinar GO Jr, Chambers KL (2015b) *Prioria dominicana* sp. nov. (Fabaceae: Caesalpinioideae), a fossil flower in mid-Tertiary Dominican amber. J Bot Res Inst Texas 9:129–134

Poinar GO Jr, Chambers KL (2015c) *Pseudhaplocricus hexandrus* gen. et sp. nov. (Commelinaceae) in mid-Tertiary Dominican amber. J Bot Res Inst Texas 9:353–359

Poinar GO Jr, Chambers KL (2016) Mimosoideae (Fabaceae) diversity and associates in mid-Tertiary Dominican amber. J Bot Res Inst Texas 10:123–136

Poinar GO Jr, Chambers KL (2021) *Salpinganthium hispaniolanum* gen. et sp. nov. (Fabaceae: Detarieae), a mid-Tertiary flowers in Dominican amber. J Bot Res Inst Texas 15:559–567

Poinar GO Jr, Columbus JT (2012) *Alarista succina* gen. et sp. nov. (Poaceae: Bambusoideae) in Dominican amber. Hist Biol 25:1–6

Poinar GO Jr, Judziewicz EJ (2005) *Pharus primuncinatus* (Poacae: Pharoideae: Phareae) from Dominican amber. Sida 21:2095–2103

Poinar GO Jr, Legalov AA (2013) *Bicalcasura maculata* n. gen., n. sp. (Curculionoidea: Dryophthoridae) with a list of weevils described from Dominican amber. Hist Biol 26:449–453

Poinar GO Jr, Mastalerz M (2000) Taphonomy of fossilized resins: determining the biostratinomy of amber. Acta Geol Hisp 35:171–182

Poinar GO Jr, Poinar R (1999) The Amber forest. Princeton University Press, Princeton, p 270

Poinar GO Jr, Santiago-Blay JA (1997) *Paleodoris lattini* gen. n., sp. n., a fossil palm bug (Hemiptera: Thaumastocoridae, Xylastodorinea) in Dominican amber, with habits discernible by comparative functional morphology. Entomol Scand 28:307–310

Poinar GO Jr, Steeves R (2013) *Virola dominicana* sp. nov. (Myristicaceae) from Dominican amber. Botany 91:530–534

Poinar GO Jr, Struwe L (2016) An asterid flower from neotropical mid-Tertiary amber. Nat Plant. https://doi.org/10.1038/NPLANTS.20165

Poinar GO Jr, Wake D (2015) *Palaeoplethodon hispaniolae* gen. n., sp. n. (Amphibia: Caudata), a fossil salamander from the Caribbean. Palaeodiversity 8:21–29

Poinar GO Jr, Chambers KL, Brown AE (2008a) *Lasiambix dominicensis* gen. and sp. nov., a eudicot flower in Dominican amber showing affinities with Fabaceae subfamily Caesalpinioideae. J Bot Res Inst Texas 2:463–471

Poinar GO Jr, Chambers KL, Brown AE (2008b) *Trochanthera lepidota* gen and sp. nov., a fossil angiosperm inflorescence in Dominican amber. J Bot Res Inst Texas 2:1167–1172

Poinar GO Jr, Weigend M, Henning T (2015) *Klaprothiopsis dyscrita* gen. et sp. nov. (Loasaceae) in mid-Tertiary Dominican amber. J Bot Res Inst Texas 9:369–379

Poinar GO Jr, Keven PG, Jackes BR (2016) Fossil species in Boehmerieae (Urticaceae) in Dominican and Mexican amber: a new genus (*Ekrixanthera*) and two new species with anemophilous pollination by explosive pollen release, and possible lepidopteran herbivory. Botany 94:599–606

Poinar HN, Cano RJ, Poinar GO Jr (1993) DNA from an extinct plant. Nature 363:677

Ramírez SR, Gravendeel B, Singer RB, Marshall CR, Pierce NE (2007) Dating the origin of the Orchidaceae from a fossil orchid with its pollinator. Nature 448(7157):1042–1045

Schlee D (1990) Das Bernstein-Kabinett. Stuttgart Beitrage für Naturkunde (C) 28:1–100

Chapter 4
Mexican Amber Flowers

Abstract Found not only in mine shafts, Mexican amber also occurs in rubble after landslides from the numerous earthquakes in the mountains around Simojovel in the state of Chiapas. The amber was and still is highly regarded, not only for its insect and plant inclusions, but for creating various art objects, especially charms for children to protect then against the evil eye (Ojo). The Lacondons, a branch of the Maya, also made amber earplugs and nose plugs from Mexican amber. Based on direct evidence, at least 14 species of angiosperms representing 16 different plant families thrived in the Mexican amber forest. Genera of Mexican amber flowers are useful in characterizing the original climate and habitat of the amber forest, which appears to be a boreotropical forest with large, fast growing trees. The original mid-Tertiary Mexican amber forest would have included the resin producing *Hymenaea,* as well as *Swietenia* mahogany trees in the canopy layer and a variety of palms and others in the subcanopy and understory. The forest floor as well as tree branches probably supported bromeliads and orchids, like the one pollinated by a small beetle.

Based on the present location of the Mexican amber mines in and around the city of Simojovel, the amber forest occupied most, if not all, of Chiapas, the southernmost state of Mexico. Within this region, the majority of the amber deposits occur in the highlands of the northern mountain ranges that support a tropical to warm-temperate climate. The earth in this area is unstable and tremors and earthquakes are frequent, some of which fracture the mountainsides and cause landslides that exposes the amber (Fig. 4.1). The amber was highly valued by the Lacondons, a branch of the Maya. They created various art objects, especially charms for children to protect then against the evil eye (Ojo) and amber earplugs and nose plugs as ornaments.

The actual amber "mines" are simple shafts chipped into the mountainsides (Fig. 4.2). The tunnels may continue for 200 or more meters with the amber occurring in coal veins surrounded by limestone (Fig. 4.3). Mining is done by various families that then sell the amber directly to buyers. Most amber mines in the Simojovel region have been dated from the Early Miocene to the Late Oligocene (22–26 mya) (Berggren and Van Couvering 1974; Vega et al. 2009a).

G. Poinar, *Flowers in Amber*, Fascinating Life Sciences, https://doi.org/10.1007/978-3-031-09044-8_4

Fig. 4.1 Landslide exposing amber in Chiapas, Mexico. (Photo taken by the late Wyatt Durham)

Thus far, all amber flowers found in Mexican amber can be placed in current families and most in extant genera. Habitats in the Mexican amber forest would have been similar to those in the Dominican amber forest in many aspects, including environments that were favorable for *Hymenaea* trees that formed the amber in both forests. While three plant genera, *Hymenaea*, *Swietenia* and *Ekrixanthera* have been described from both Mexican and Dominican amber, their corresponding species differ and up to the present there are no common angiosperm species found in these two amber deposits.

Hymenaea mexicana, description based on Poinar and Brown 2002

Mexican amber is formed from resin of *Hymenaea mexicana* trees. The flowers of both *Hymenaea protera* and *Hymenaea mexicana* have 3 large petals that fall off soon after opening and 2 small persistent petals. This is one of the characters that links both of the extinct amber *Hymenaea* species to the living East African

Fig. 4.2 Entrance to amber mine in Chiapas, Mexico

Fig. 4.3 Mine through limestone with amber occurring in coal veins

Hymenaea verrucosa. There are slight floral differences between *Hymenaea mexicana* and the Dominican amber *Hymenaea protera*. The size of the basal midrib and the degree of extension of the basal lobes of the lamina of the large petals differs in the two species. The 3 large showy petals of *Hymenaea mexicana* also have thicker stalks and the blades are more rounded above (Fig. 4.4).

It is possible that all three *Hymenaea* species evolved from an ancestral lineage in the South America- African landmass. Then, as the Greater Antilles separated from Central America, the Mexican and Dominican species began to differ based on edaphic, biological and climatic factors. While they flourished in the tropical, moist Dominican and Mexican amber forests, neither species is found today in the New or Old Worlds. The same situation occurrs with the animals found in amber from both areas. No extant species, based on morphology, have been recovered from either amber source.

A second species of *Hymenaea* was described from Mexican amber. This species (*Hymenaea allendis* Calvillo-Canadell et al. 2010) is characterized by a prominent nectariferous disc and smooth glabrous ovary. Otherwise, it is similar to *Hymenaea mexicana*.

Colpothrinax chiapensis, description based on Chambers et al. 2012

These hermaphrodite palm flowers in the family Arecaceae are quite striking even though they only are 6 mm in diameter. Several flowers were together in a

Fig. 4.4 Large petal of *Hymenaea mexicana*

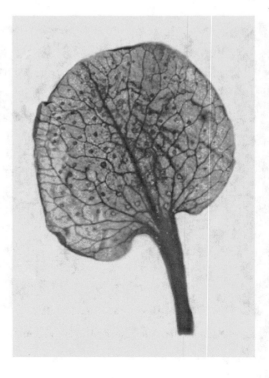

single amber piece, indicating that they came from a floral cluster. There are 3 small, cup-shaped sepals, 3 extended petals, 6 stamens and straight style arising from the trifid ovary (Figs. 4.5 and 4.6). The anthers in some flowers are missing and could have been removed by herbivores or shed after releasing their pollen (Fig. 4.5). Perhaps the former since in other flowers that appear to be in the same stage of development, the anthers are present (Figs. 4.6 and 4.7).

While many palms are thought to be wind pollinated, some may also be insect pollinated since bees, leafhoppers and bruchid beetles visit extant palm flowers. In fact, the shiny, white eggs that are deposited on the petal of *Colpothrinax chiapensis* not far from the developing ovary resemble those of palm bruchids (Fig. 4.6). After hatching, the bruchid larvae enter the pistil and molt into sedentary grubs that complete their development on the palm eggs. When mature, the grubs pupate in the empty seeds and emerges as adults.

It is thought that beetles represent one of the earliest known general insect pollinators of flowers (Bernhardt 2000) and there are many record of beetles visiting flowers today (Li et al. 2021). The palm seeds that escape the bruchids mature inside a fleshy fruit that is eaten and distributed by birds, mammals and even fish.

There are no *Colpothrinax* palms in Mexico today but they do occur in Central America and the Caribbean. Based on current species of *Colpothrinax*, the habitat of *Colpothrinax chiapensis* was that of a moist tropical or subtropical forest. These fan palms, with leaves clustered near the top of the stems, are used today in

Fig. 4.5 *Colpothrinax chiapensis* flower

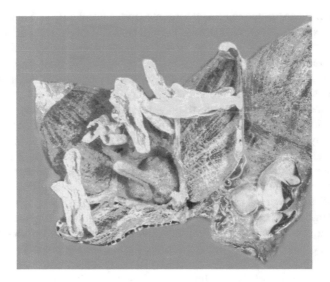

Fig. 4.6 *Colpothrinax chiapensis* with insect eggs attached to petal

Fig. 4.7 *Colpothrinax chiapensis* flower with anthers and adjacent leafhopper

landscaping. The tree provides an interesting silhouette, especially those species with swollen trunks.

Socratea brownii, description based on Poinar 2002

A second palm species in Mexican amber is a multi-stamen palm of the genus *Socratea*.

Typical for male flowers of the genus, *Socratea brownii* possesses numerous (50-some) stamens. While 3 sepals, 3 petals and 6 stamens are normal features of palm flowers, only a few living male palm flowers have up to several hundred stamens.

The sepals of *Socratea brownii* are greatly reduced but the slightly asymmetrical petals are distinct and fleshy. The numerous stamens fill the center of the flower and since the filaments are so short, only the anthers are visible (Fig. 4.8). It is probably the pollen that attracted the springtails that were adjacent to the blooms (Fig. 4.9).

Socratea palms are absent from Mexico today and the five living species occur in Central and South America. These palms are noted for their adventitious stilt roots that emerge near the base of the stem and enter the ground, thus providing support for the stalk. Local inhabitants claim that these trees can "walk". Apparently they can move horizontally over time by growing stilt roots on one side and relinquishing them on the opposite side. Toucons eat the bulky fruits (drupes) and spiny pocket mice feed on the seeds.

Fig. 4.8 Two *Socratea brownii* palm flowers

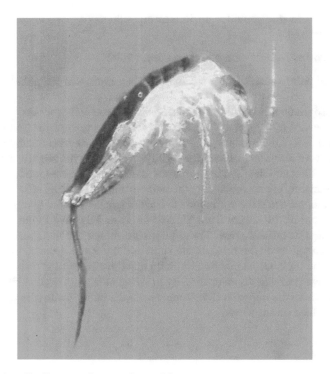

Fig. 4.9 Springtail adjacent to *Socratea brownii* flower

Ekrixanthera ehecatli, description based on Poinar et al. 2016

These little pentamerous male flowers of the nettle family (Urticaceae) range between 3 and 4 mm in width. *Ekrixanthera ehecatli* has 5 sepals and 5 stamens but petals are absent. In the center of the flower is a small sterile pistilloid (Fig. 4.10).

One interesting feature of members of this genus is what is known as "explosive pollen release". When the pollen in the anther is mature, it discharges in an explosive outburst, thrusting the grains into the air where the wind will carry them to receptive female flowers. Many flowers in this family with "explosive pollen release" live on the forest floor along stream banks, where the water currents generate faint air movements that assist in carrying the pollen grains. Falling in the resin apparently triggered such an explosion in one of the anthers, leaving the ejected pollen grains forever entombed adjacent to the anther (Fig. 4.11). This type of pollen release indicates that the female flowers were wind pollinated, which explains why no nectar secreting tissues to attract insects could be found on these male flowers. A tan-colored paper for writings and drawings known as amate is still fabricated today by Lacondonian Indians from bark fibers of *Ekrixanthera* spp.

Fig. 4.10   Flower of *Ekrixanthera ehecatli*

Fig. 4.11   Explosive pollen release in *Ekrixanthera ehecatli*

Gouania miocenica Hernández-Hernández and Castañeda-Posadas 2018

Gouania miocenica was described from Mexican amber by Hernández-Hernández and Castañeda-Posadas (2018). The small flowers of members of this genus are pentamerous with triangular sepals and hooded, round tipped petals (Fig. 4.12). The curious feature of the flower, as well as many other members of the Rhamnaceae family, are how the petals, which are situated opposite the stamens, tend to enclose the basal portions of the stamens (Fig. 4.13), similar to the Dominican amber flower, *Comopellis presbya*. Present day *Gouania* species are shrubs and lianas that thrive in tropical- subtropical climates, which were the conditions of the Mexican amber forest. The nectar-bearing disc attracts a number of insects, even mosquitoes, but the majority of pollinators appear to be bees. The fleshy fruits are eaten by vertebrates that then disperse the seeds.

***Acacia* sp.**, description from Miranda 1963

This little *Acacia* flower that shows herbivore damage to its 5 sepals, 5 petals, 10 stamens and erect pistil (Fig. 4.14) may belong to the same *Acacia* sp. previously described by Miranda (1963) that was based on a partial leaf in Mexican amber.

Actually mimosoids (which include acacias) are one of the most successful plant groups today. Fossils tell us that populations of mimosoids existed some 40 million years ago in Europe, Africa and North America, with the densest regions in tropical America and Africa. The flowers are relished by many insect herbivores, including

Fig. 4.12 Flower of *Gouania miocenica*

Fig. 4.13 Petal and stamen of *Gouania miocenica*

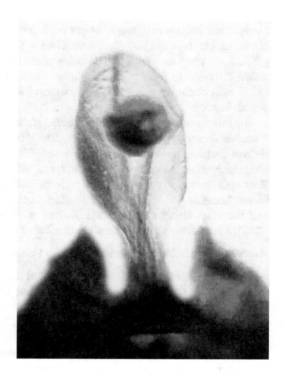

Fig. 4.14 *Acacia* sp. flower

caterpillars, beetles and thrips. Major pollinators appear to be bees, which come to obtain nectar from the tiny glands attached to the anthers.

Ants are frequent visitors to both the petals and leaves of mimosoids. Some ants guard the trees to protect nectar glands on the leaf petioles that they visit daily. Other ants come to search for prey, which include noctuid moth caterpillars and scarab beetles.

An interesting example of this symbiotic association involves ants of the genus *Pseudomyrmex* that live inside the swollen thorns of certain species of acacia and defend the plant against various herbivores. Birds, like the little rufous-naped wren, commonly make their nests in acacia trees, using the ants to keep predators away. Bats visit the trees to feed on the ripe fruits and then void the seeds throughout the forest. Larvae of bruchine beetles develop on acacia seeds while still inside the pods and the adult beetles feed on the flowers of plants not patrolled by ants (Jansen 1983). Leaf-infecting fungi may be distributed by various insect herbivores.

Annulites mexicana description based on Poinar 2016

As noted in Baltic and Dominican amber, the presence of orchids can often be determined by pollen sacs (pollinaria) attached to various insects that have visited the flowers. In this case, the beetle is quite small (1.4 mm long) and the pollinarium is attached to its head (Fig. 4.15).

The elongate strongly annulated pollinia of *Annulites mexicana* are 0.5 mm long (Fig. 4.16) and contains transverse rows of spherical pollen tetrads. It would be interesting to know how the pollinia become detached when the beetle enters another orchid flower. Some present day orchids in the tropical American genus *Solenocentrum* possess similar pollinia. These terrestrial, small to medium sized orchids have a trailing habit and bear terminal clusters of whitish flowers.

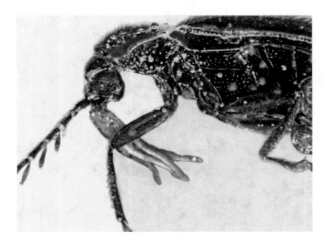

Fig. 4.15 *Annulites mexicana* with pollinarium

Fig. 4.16 Tips of attached pollinia of *Annulites mexicana*

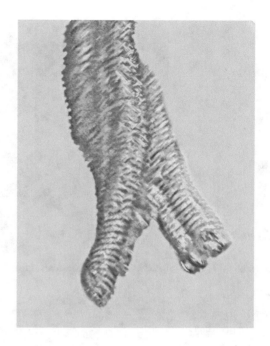

The transporter of the pollinarium is a member of the family Ptilodactylidae, a group of small beetles that favor semi-aquatic habitats like swamps and marshes. It was probably searching for nectar when it encountered the orchid pollinarium.

Podopterus sp.

Isolated fruits and seeds frequently occur in amber but it is not easy to identify the plant source. A pair of attractive tri- winged fruits (15 mm and 22 mm in length) in Mexican amber (Fig. 4.17) were identified by Pedro Acevedo-Rodriguez as belonging to the genus *Podopterus* of the family Polygonaceae. This genus contains several species that occur today in Mexico and Central America. The single Mexican species, *Podopterus mexicanus*, is a small tree found in two separate habitats, arid thorn forests in Mexico and tropical evergreen forests in Costa Rica (Jansen 1983). It is strange to have the same species adapted to such different habitats. While we cannot speculate from which of these environments the fossil tri-winged fruit originated, nevertheless, it shows that this genus was present in the Mexican amber forest.

4.1   Indirect Evidence of Plant Families

While there is no direct evidence of some plant families in Mexican amber based on flowers, their existence can be surmised by the present of other organisms that are dependant on them. One example is the presence of a crab in Mexican amber

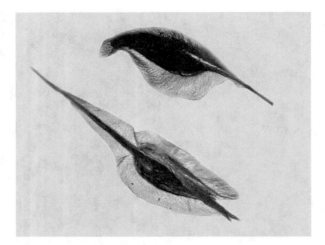

Fig. 4.17 Fruits of *Podopterus* sp. in Mexican amber

Fig. 4.18 Crab in Mexican amber that could indicate presence of tank bromeliads

(Fig. 4.18). This crab was identified as a member of the genus *Sesarma* of the family Sesarmidae (Vega et al. 2009b). This genus is restricted to the New World tropics and at least one extant species lives in water collected at the base of bromeliad leaves. It is possible that the fossil crab had a similar habitat, thus indicating the presence of bromeliads in the Mexican amber forest.

4.2   Summary of Mexican Amber Flowers

Based on direct and indirect evidence, representatives of 16 different plant families thrived in the Mexican amber forest (Table 4.1). Some of these families, such as the Fabaceae, Orchidaceae, Bromeliaceae, Arecaceae, Urticaceae and Rhamnaceae are also represented in Dominican amber. However there are no records of the Anacardiaceae, Araliaceae, Ericaceae, Polygonaceae, Salicaceae and Staphyleaceae in Dominican amber.   While a few genera (*Hymenaea, Ekrixanthera Swietenia*) are found in both amber deposits, there are no common species amongst them. While the absence of common species may be due to the paucity of Mexican amber flowers, it could also indicate some major habitat and age differences between the two fossil locations.

Although 5 out of the 38 flowers in Dominican amber belong to unknown families, all of the Mexican amber flowers can be placed in current families. However, it is too early to draw any conclusions about Mexican amber flowers being younger those in Dominican amber. None of the Mexican or Dominican amber flowers can be assigned to present day species.

Table 4.1 Mexican amber flowers (N = 14)

Plant	Systematic placement	References
Acacia sp.	Fabaceae	Miranda (1963); present work
Annulites mexicana	Orchidaceae	Poinar (2016)
Aralia sp.	Araliaceae	Hernández-Hernández and Castañeda-Posadas (2020)
[a]tank bromeliad	Bromeliaceae	Vega et al. (2009a, b)
Colpothrinax chiapensis	Arecaceae	Chambers et al. (2012)
Ekrixanthera ehecatli	Urticaceae	Poinar et al. (2016)
Gouania miocenica	Rhamnaceae	Hernández-Hernández and Castañeda-Posadas (2018); present study
Hymenaea allendis	Fabaceae	Calvillo-Canadell et al. (2010)
Hymenaea mexicana	Fabaceae	Poinar and Brown (2002)
Lunania floresi	Salicaceae	Hernández-Damián et al. (2016)
Podopterus sp.	Polygonaceae	Present work
Socratea brownii	Arecaceae	Poinar (2002)
Staphylea ochoterenae	Staphyleaceae	Hernández-Damián et al. (2018)
Swietenia miocenica	Meliaceae	Castañeda-Posadas and Cevallos-Ferriz (2007)
Tapirira durhamii	Anacardiaceae	Miranda (1963)
Unknown genus	Ericaceae	de Hernández-Hernández et al. (2020)
Unknown genus	Santalaceae	Hernández-Hernández et al. (2020)

[a]Based on indirect evidence from amber crab

The original mid-Tertiary Mexican amber forest probably had a distinct canopy layer, at least during part of its existence, composed of *Hymenaea mexicana, Hymenaea allendis* and *Swietenia miocenica*. The subcanopy and understory would have been represented by *Aralia* sp., *Socratea brownii, Tapirira durhamii* and *Podopterus sp,* The shrub layer could have included *Colpothrinax chiapensis,* as well as acacias and representatives of the Ericaceae. The florest floor as well as tree branches could have supported bromeliads and orchids such as *Annulites mexicana.*

Genera of Mexican amber flowers are usuful in characterizing the original climate and habitat of amber forest (Hernández-Hernández and Castañeda-Posadas 2020). For instance, the flower of *Tapirira durhamii* suggests a variable climate since some populations of the related *Tapiriria mexicana* occur today in highly humid and warm environments, while other populations are found in high altitude forests with moderately cool climates (Miranda 1963).

Based on its present day descendants, the small (4 mm long) bisexual flower of *Lunania* of the willow family (Hernández-Damián et al. 2016) was one of many on a long stem descending from the leaf bases of a small tree in the Mexican amber forest. While mention of the family Salicaceae brings to mind willow trees that are common in the temperate climates of North America, the family is now quite diverse and contains lineages that grow under tropical conditions. *Lunania* is one of these genetic lines with all 5 living species growing in subtropical- tropical forests in Cuba and Jamaica (Hernández-Damián et al. 2016).

The relatively large (11 mm long) bisexual flower of *Staphylea* (Staphyleaceae) described from Mexican amber (Hernández-Damián et al. 2018) was probably a shrub or small tree based on the habits of its present day descendants. Today, the genus contains some 10–13 species that grow in subtropical to temperate habitats. Today, in forests in the west-central state of Guerrero, Mexico, *Staphylea, Hymeneae* and *Lunania* all grow together in what is known as a Boreotropical forest, which could very well typify the habitat of the Mexican amber forest during most of its existence.

Key to Mexican Amber Flowers

1. Isolated mature tri-winged fruits – *Podopterus* sp.

    1A. Flowers or floral parts – 2

2. Elongate pollinarium attached to mouthparts of beetle – *Annulites mexicana*

    2A. Flowers not represented by pollinaria – 3

3. Flowers lacking petals (or with only 1 whorl of tepals) – *Lunania floresi*

    3A. Flowers with petals (or tepals) – 4

4. Flowers with 3 petals (or tepals) – 5

    4A. Flowers with 5 petals (or tepals) – 6

5. Flowers with some 50 stamens – *Socratea brownii*

    5A. Flowers with 6 stamens – *Colpothrinax chiapensis*

6. Flowers with 3 large and 2 reduced petals – 7

   6A. Flowers with similar size petals (or tepals) – 8

7. Flowers with prominent necteriferous disc and glabrous ovary – *Hymenaea allende*

   7A. Flowers with small necteriferous disc and hirsute pubescence at base of ovary –*Hymenaea mexicana*

8. Flowers with 10 stamens – 9

   8A. Flowers with 5 stamens – 10

9. Flower male, with a 5 lobed pistillode with 5 small styles – *Tapirira durhamii*

   9A. Flower with functional pistil – *Acacia* sp.

10. Base of stamens opposite to and enclosed by petals – *Gouania miocenica*

   10A. Flowers not as above – 11

11. Flowers wide, spreading; explosive pollen release – *Ekrixanthera ehecatli*

   11A. Flowers bell-shaped; no explosive pollen release – *Staphylea* ochoterenae

References

Berggren WA, Van Couvering JAH (1974) The late Neogene: biostratigraphy, geochronology and paleoclimatology of the last 15 million years in marine and continental sequences. Palaeogeogr Palaeoclimatol Palaeoecol 16:1–216

Bernhardt P (2000) Convergent evolution and adaptive radiation of beetle-pollinated angiosperms. Plant Syst Evol 222:293–320

Calvillo-Canadell L, Cevallos-Ferriz SRS, Rico-Arce L (2010) Miocene *Hymenaea* flowers preserved in amber from Simojovel de Allende, Chiapas, Mexico. Rev Palaeobot Palynol 160:126–134. https://doi.org/10.1016/j.revpalbo.2010.02.007

Castañeda-Posadas AC, Cevallos-Ferriz SRS (2007) *Swietenia* (Meliaceae) flower in Late Oligocene-Early Miocene amber from Simojovel de Allende, Chiapas, Mexico. Am J Bot 94:1821–1827

Chambers KL, Poinar GO Jr, Brown AE (2012) A new fossil species of *Colpothrinax* (Arecaceae) from Mid-Tertiary Mexican amber. J Bot Res Inst Texas 6:557–560

de Hernández-Hernández MJ, Castañeda-Posadas C (2018) *Gouania miocenica* sp. nov. (Rhamnaceae), a Miocene fossil from Chiapas, México and paleobiological involvement. J S Am Earth Sci 85:1–5

de Hernández-Hernández MJ, Castañeda-Posadas C (2020) Paleoclimatic and vegetation reconstruction of the Miocene Southern Mexico using fossil flowers. J S Am Earth Sci 104. https://doi.org/10.1016/j.jsames.2020.102827

Hernández-Damián AL, Cevallos-Ferriz SRS, Canadell LC (2016) Flower of a new species of *Lunania* Hook (Salicaceae sensu lato-Samydeae) as inclusions in Miocene amber from Simojovel de Allende, Chiapas, Mexico. Bol Soc Geol Mex 68:29–36

Hernández-Damián AL, Cevallos-Ferriz SRS, Huerta-Vergara AR (2018) Fossil flower of
    Staphylea L. from the Miocene amber of Mexico: new evidence of the Boreotropical Flora in
    low-latitude North America. Earth Environ Sci Trans R Soc Edinburgh 108:471–478
Jansen DH (1983) Costa Rican natural history. University of Chicago Press, Chicago. 816 pp
Li K-Q, Ren Z-X, Li Q (2021) Diversity of flower visiting beetles at higher elevations on the
    Yulong Snow Mountain (Yunnan, China). Diversity 13. https://doi.org/10.3390/d13110604
Miranda F (1963) Two plants from the amber of the Simojovel, Chiapas, Mexico, area. J Palentol
    37:611–614
Poinar GO Jr (2002) Fossil palm flowers in Dominican and Mexican amber. Bot J Linnean Soc
    138:57–61
Poinar GO Jr (2016) Beetles with orchid pollinaria in Dominican and Mexican amber. Am Entomol
    62:180–185
Poinar GO Jr, Brown AE (2002) *Hymenaea mexicana* sp. nov. (Leguminosae: Caesalpinioideae)
    from Mexican amber indicates Old World connections. Bot J Linnean Soc 139:125–132
Poinar GO Jr, Keven PG, Jackes BR (2016) Fossil species in Boehmerieae (Urticaceae) in
    Dominican and Mexican amber: a new genus (*Ekrixanthera*) and two new species with ane-
    mophilous pollination by explosive pollen release, and possible lepidopteran herbivory. Botany
    94:599–606
Vega FJ, Nyborg T, Coutiño MA, Solé J, Hernández-Monzón O (2009a) Neogene Crustacea from
    Southeastern Mexico. Bull Mizunami Foss Mus 35:51–69
Vega FJ, Zúñiga L, Pimentel F (2009b) First formal report of a crab in amber from the Miocene
    of Chiapas and other uncommon Crustacea. Geological Society of America annual Meeting,
    Portland, OR, 18–21 October 2009, paper no. 245 20:1–3

General Conclusions

This survey of amber flowers presents snapshots of floral development from the mid-Cretaceous to the mid-Tertiary. It is obvious that the majority of mid-Cretaceous flowers are genetic stalemates that disappeared over time. Only a few flower families, such as the Poaceae and Lauraceae, can be linked to present day families.

By the mid-Tertiary, most flowers can be placed in modern families and many in known genera. However, even then, a few flower lineages did not survive even at the family level. There are numerous reasons for their extinctions, among them being competition with other floral lineages, habitat or environmental change, absence of pollinators, foliar and floral herbivores and plant pathogens (especially fungi). All of these can produce genetic bottlenecks, causing extinctions of population or entire lineages.

Along with fine cellular details of flowers, amber also preserves associated pollinators and herbivores. Not only are these arthropods vital in the survival of angiosperms, but their size and biology (based on descendants) reveal clues about ancient microhabitats.

Theories on the origin and evolution of angiosperms have been presented, discussed and refuted since Charles Darwin referred to their presence as an "abominable mystery" (Doyle 1978; Soltis et al. 2019; Bateman 2020; Berendse and Scheffer 2009; Cascales-Minana et al. 2016; Magallón and Castillo 2009). Even whether angiosperms are monophyletic or phylogenetic is undecided. It is curious why none of the above named authors discuss flowers in Burmese amber. Especially since this amber source is dated from the mid-Cretaceous, which is soon after the first undisputed fossil records of angiosperms appeared in the Early Cretaceous (Friis and Endress 1990; Crane et al. 2004; Friis et al. 2011). This is one of the rare sites where angiosperm diversification is clearly evident.

Since flowers in Burmese amber originated from Gondwana (Poinar 2018), it is possible that Western Gondwana was the birthplace of angiosperms some 120 mya. Presently, Australian rainforests contain the greatest concentration of primitive flowering plants in the world (White 1986, 1994). The progenitors of these early

G. Poinar, *Flowers in Amber*, Fascinating Life Sciences, https://doi.org/10.1007/978-3-031-09044-8

angiosperm lineages then radiated out with landmasses that formed Gondwana, especially South America, Africa and India. Australia appears to be unique by not only having the most primitive angiosperms, but also being the home of the most primitive vertebrates (monotremes).

References

Bateman RM (2020) Hunting the Snark: the flawed search for mythical Jurassic angiosperms. J Exp Bot 71:22–35

Berendse F, Scheffer M (2009) The angiosperm radiation revisited, an ecological explanation for Darwin's 'abominable mystery'. Ecol Lett 12:865–872

Cascales-Minana B, Cleal CJ, Gerrienne P (2016) Is Darwin's 'abominable mystery' still a mystery today? Cretac Res 61:256–262

Crane PR, Herendeen PS, Friis M (2004) Fossils and plant phylogeny. Am J Bot 91:1683–1699

Doyle JA (1978) Origin of angiosperms. Annu Rev Ecol Syst 9:365–392

Friis EM, Endress PK (1990) Origin and evolution of angiosperm flowers. Adv Bot Res 17:99–162

Friis EM, Crane PR, Pedersen KR (2011) Early flowers and Angiosperm evolution. Cambridge University Press, Cambridge, 585 pp

Magallón S, Castillo A (2009) Angiosperm diversification through time. Am J Bot 96:349–365

Poinar GO Jr (2018) Burmese amber: evidence of Gondwanan origin and Cretaceous dispersion. Hist Biol 31:1304–1309

Soltis PS, Folk RA, Soltis DE (2019) Darwin review: angiosperm phylogeny and evolutionary radiations. Proc R Soc B 286:20190099. https://doi.org/10.1098/rspb.2019.0099

White ME (1986) The greening of Gondwana. Reed Books, Frenchs Forest, 356 pp

White ME (1994) After the greening; the browning of Australia. Kangaroo Press, Kenthurst, 288 pp

Index

A

Acacia sp., 200, 201, 205, 207
Actinobacterium, 50, 52
Agaonidae, 178, 184
Agathis, 2–57, 71, 84
Alarista succina, 147, 149, 150, 184, 185
Amber, v, vii, ix, xiii, 2, 3, 6, 12, 27, 49, 52,
    69–72, 75, 77, 81, 84, 95, 96, 98, 99,
    109, 113, 115–121, 124, 126, 154–156,
    158, 162, 168, 171, 178, 191–195, 203,
    205, 206, 209
Amber mines, xiii, 97, 191, 193
Anacardiaceae, 205
Angiosperms, vii, xi, xiii, 1, 7, 14, 16, 22, 50,
    53, 54, 57, 60, 62, 69, 82, 93, 95, 192,
    209, 210
Annulites mexicana, 202, 203, 205, 206
Ant, 117, 118, 127, 130, 137, 138, 144,
    147, 150
Anther glands, 113, 118
Antiquifloria latifibris, 27, 29, 37, 61, 64
Apocynaceae, 119, 184
Aralia sp., 205, 206
Araliaceae, 205
Araucariaceae, vii, 2, 71
Arecaceae, 93, 171–178, 184, 185, 194, 205
Atherospermataceae, 15, 61
Australia, xiii, 2, 35, 62, 73, 210

B

Baltic amber, vii, 69–93
Baltic amber forest, vii, 70, 71, 76, 77
Bamboo, 14, 150, 170

Bicalcasura maculata, 175
Bignoniaceae, 184
Brevitrimaris arcuatus, 100, 184, 185
Bromeliaceae, 178, 184, 205
Burma (Myanmar), vii, 2
Burmese amber, vii, xi, 1–65, 209
Burmese amber forest, 1–3, 10, 14, 26,
    34, 60, 62
Burseraceae, 153, 184

C

Camouflaged beetles, 3, 7
Campopetala dominicana, 113, 114, 184, 186
Carpantholithes berendtii, 79, 80, 87–89, 93
Cascolaurus burmitis, 34, 35, 38, 39, 61, 64
Cascomixticus tubuliferous, 47
Cascoplecia insolitis, 35
Catalpa hispaniolae, 153, 184, 187
Caterpillars, 78, 79, 86, 117, 130, 137,
    154, 202
Cecidomyiidae, 118, 158
Cedar twigs, 75, 77
Celastraceae, 146, 184
Cephalotus ant, 165
Chainandra zeugostylus, 42, 44, 61, 64
Chenocybus allodapus, 53, 55, 61, 63
Chiapas, xiii, 191–193
Chrysobalanaceae, 166, 184
Cladarastega burmanica, 6
Clethraceae, 80, 93
Coccoliths, 96
Colpothrinax chiapensis, 194–196, 205, 206
Commelinaceae, 112, 184

Printed in the United States
by Baker & Taylor Publisher Services